1 MONTH OF
FREE
READING

at

www.ForgottenBooks.com

By purchasing this book you are eligible for one month membership to ForgottenBooks.com, giving you unlimited access to our entire collection of over 1,000,000 titles via our web site and mobile apps.

To claim your free month visit:

www.forgottenbooks.com/free854113

ISBN 978-0-267-51105-1
PIBN 10854113

NOTES

ON

MECHANICS

BY

CHARLES R. CROSS

Printed for the use of Students

IN THE

Massachusetts Institute of Technology

1908

BOSTON

WM. B. LIBBY, THE GARDEN PRESS

16 ARLINGTON STREET

Copyright, 1908

By CHARLES R. CROSS

MECHANICS.

1. Motion.—Motion is change of position in space. We can be acquainted with none but relative motions, as we can know that a body really changes its position only by comparison with some other body not possessing the same movement. Our use of the term "rest" is also relative.

In pure motion of translation all the points of a body move with the same velocity and in the same direction at the same instant. When the points of a body describe arcs of concentric circles about some fixed axis, the motion is one of pure rotation. All possible varieties of motion may be produced by the combination of translation and rotation.

2. Velocity.—As the term is commonly used, velocity is speed, or rate of motion. It is expressed analytically by the formula $v = \dfrac{ds}{dt}$. For uniform motion it is evident that speed equals the space traversed in any given time divided by that time.

In physical treatises, however, it is usual to distinguish between velocity and speed, making the former a vector quantity, involving both magnitude and direction, while the latter denotes magnitude only, as defined above.

It is furthermore evident that the speed of a particle at any instant is always the space which it would describe in a unit of time were its speed to remain of the same value during that time.

It is customary to express speed in terms of the number of units of length which would be thus traversed in a second; e. g., a freely falling body at the end of its first second of fall has a speed of 9.8 meters per second.

3. Force.—We have come to ascribe every change that occurs in the condition or qualities of matter to the action of force. Force is commonly defined as that which causes or tends to cause a change of condition. A mechanical force, or pondero-motive force as it is sometimes called, tends to produce motion of a mass.

An electro-motive force, on the other hand, tends to produce a flow of electricity. A magneto-motive force tends to produce magnetic flux. But these are not forces in a mechanical sense.

It will readily be seen that what we call "force" is really known to us only as a conception. All that we learn directly from experience is that given certain conditions certain results invariably follow. We conceive of something which we call force as acting to bring about these results.

Thus when a stone is raised above the surface of the earth it tends to fall to the earth, and will do so if unsupported, while if supported it exerts a pressure upon the support. This tendency we ascribe to a force which we call "gravity" acting between the stone and the earth. But all that we know experimentally is that the stone and earth tend to approach each other, and that this tendency has a definite magnitude in any particular case.

Also, when we speak of a force as acting for a certain time, the only fact that we know from observation is that the conditions under which certain actions take place persist for a certain time.

4. Composition of Velocities and Forces.—We shall for the present consider only the case in which these are applied at a point.

Velocities and forces are vector quantities. Hence the combined effect or resultant of a number of velocities or forces can be found by the usual process of vector addition. This is the case whether the components lie in the same or in different planes.

Applications of this principle are the familiar propositions known as the Parallelogram and Triangle of Velocities, the Parallelogram and Triangle of Forces, the Polygon of Forces, the Parallelopiped of Forces.

5. Equilibrium of Forces.—It follows from the facts stated in the preceding paragraph that two forces are in equilibrium when they are equal and opposite, since their resultant is then equal to zero. Also for a like reason any number of forces are in equilibrium when they can be represented by all the sides of a polygon taken in order.

6. Couple.—A moment (*torque*) or a couple (see § 29, p. 9) is also a vector quantity, since it acts to produce pure rotation in a definite plane. It is customary to represent a moment (*force × arm*) by a distance laid off on the axis, right-handed rotations being made positive. Couples may be combined by ordinary vector addition, as will be explained later.

7. Pressure and Impulse.—An impulse is to be considered merely as a pressure acting for a very short time. Any pressure may be considered as due to a series of recurring impulses, separated by very short intervals. Such, for example, is the explanation of gaseous pressure offered by the Kinetic Theory.

8. Mass.—The term "mass" is commonly used to denote the quantity of matter that a body contains.

9. Measure of Forces.—We shall shortly see that all mechanical forces may be compared by their effects in producing or modifying motion.

Steady forces may be measured statically by the use of various forms of dynamometer, of which the ordinary spring balance is an example.

10. Acceleration.—Acceleration is rate of change of velocity. It is expressed by the equation $a = \dfrac{dv}{dt} = \dfrac{d^2s}{dt^2}$.

It is evident that if the acceleration is constant,—that is, if equal increments of velocity are gained in successive equal times, the acceleration is equal to the velocity gained in a unit of time.

Thus the acceleration of a body falling freely under the influence of its weight is 9.8 meters per second per second; often expressed as 9.8 meter / (second)². As the unit of time habitually used in physical measurements is the second, this may be more briefly stated as 9.8 meters, the expression "per second per second" being understood.

11. Mechanics.—Mechanics, in the sense in which the term is ordinarily used, is that branch of physics which treats of the action of force on bodies. It is commonly divided into Statics and Dynamics. Statics treats of balanced forces, or forces in équilibrium; Dynamics, of the action of forces in producing motion. A better nomenclature, however, is that used by Thomson and Tait, as shown in the following quotation from their "Elements of Natural Philosophy" (1873):—

"The science which investigates the action of force is called by the most logical writers Dynamics. It is commonly but erroneously called Mechanics; a term employed by Newton in its true sense, the Science of Machines and the Art of making them.

"Force is recognized as acting in two ways: (1) so as to compel rest or to prevent change of motion; and (2) so as to produce or to change motion. Dynamics, therefore, is divided into two parts which are conveniently called Statics and Kinetics."

THREE LAWS OF MOTION.

The following propositions, first clearly and collectively stated by Newton, are shown to be true by universal experience. The paragraphs in quotation marks are given in translation as originally stated by Newton in the "Principia," published in 1687.

12. Law I.—"*Every body continues in its state of rest, or of uniform motion in a straight line, except in so far as it may be compelled by force to change that state.*"

Another mode of stating the law is the following: *A body at rest continues in that state, and a body in motion proceeds uniformly in a straight line, unless acted upon by some external force.* This truth is the principle of the *Inertia of Matter.*

ILLUSTRATIONS OF LAW I.: Railroad accidents; phenomena observed by standing passenger on Boston and other street railway cars; coursing; fixing head of hammer.

13. Law II.—"*Change of motion is proportional to the force applied, and takes place in the direction in which that force acts.*" The term "change of motion," as used by Newton, is identical with the term "change of momentum," now universally employed.

14. Momentum.—The product MV of a mass M by its velocity V is called its momentum, which is evidently proportional to both M and V. If we call a the acceleration produced by any force F acting upon a mass M, the rate of change of momentum will be the mass multiplied by the rate of change of its velocity, or Ma. Hence the following statement is true: *Forces are proportional to the momenta which would be generated by their constant and uniform action during a unit of time, or equal times;* or, in other words, *the unbalanced force acting upon a body is measured by the rate of change of momentum thereby produced.* Algebraically this is indicated by the expression $F \propto Ma$, in which F is the force producing an acceleration a in the mass M.

If the force F acts in opposition to a motion already existing in the body, a may be regarded as negative.

If we suppose the same force F to act on different masses M_1, M_2, producing accelerations a_1, a_2, respectively, we shall have $F \propto M_1 a_1$, $F \propto M_2 a_2$, whence $M_1 a_1 = M_2 a_2$ and $a_1 : a_2 :: M_2 : M_1$. Hence, *The velocities impressed upon different masses by the action of equal forces during a unit of time, or during equal times, are inversely proportional to those masses.*

The following propositions are derived immediately from the relation $F \propto Ma$.

If M is constant, F is proportional to a; whence, *Forces are proportional to the accelerations which they impress upon equal masses.*

If a is constant, F varies as M, whence, *Forces are proportional to the masses upon which they impress equal accelerations.*

It will be shown later that the momentum MV generated by the action of a force F on a mass M for a time t is equal to the magnitude of the force multiplied by its duration; *i.e.*, $Ft = MV$. The quantity Ft is called the *impulse* of the force. Hence the effect of a force is measured by its magnitude multiplied by the time during which it acts.

15. Corollary to Law II.—*When any number of forces act simultaneously upon a body, then, whether the body be originally at rest or in motion, each force produces exactly the same effect in magnitude and direction as if it acted alone.* This principle is often called the *Law of the Independence of Motions.*

ILLUSTRATIONS: Man walking or writing on vessel; Parry's sailors on ice-floe; body falling from an elevation; cannon-ball fired in different directions with regard to motion of earth.

16. Law III.—"*To every action there is always an equal and contrary reaction; or, the mutual actions of two bodies are always equal and oppositely directed.*"

ILLUSTRATIONS: Attraction or repulsion of magnets; action of uncoiling spring; recoil of gun; mutual attraction of earth and mass.

17. Time Required to Produce Motion of a Mass.—When a force is applied to a large mass the mass necessarily acquires velocity only with comparative slowness; that is, the acceleration is small. The greater the mass, with a given force, the more slowly does the mass gain in speed. Hence if the force is applied to the mass, for example, by means of a spring or cord, the more sudden the application of the force, the more will the spring or cord be stretched. The mass seems to offer a resistance to entering into motion, though it is nevertheless acquiring velocity at the rate conditioned by the 2d Law of Motion; *i.e.*, $a = \dfrac{F}{M}$.

This effect is often said to be due to the "time required to overcome the inertia of a body." Such a statement is erroneous if the word "inertia" is used in its strict sense; *viz.*, as denoting the absolute inability of matter to change its state except under the action of force. But this term is very commonly used in a somewhat different though analogous sense as denoting that property of matter in virtue of which a definite force is necessary to produce a given change in the existing state of a mass. This last is simply a mode of expressing the general idea of momentum. The same idea is often expressed by the statement that "time is required to produce motion in a mass."

Many peculiar phenomena are explained in virtue of this action; as, for example, the feat of firing a candle through a board, the breaking of the harness under the sudden pull of a horse, the bursting of cannon with high-power explosives, surface-blasting with dynamite, and the like.

The three Laws of Motion, as is the case with all physical principles, are known to be true only from observation and experiment. The best proof that we have of their universality is found in the accordance of observed with predicted results, as for example in Astronomy in the case of eclipses, occultations, planetary and cometary motions and other phenomena in which the mutual actions of masses of matter are concerned.

ANALYTICAL STATICS.

18. Analytical Statics.—The following elementary applications of the principles of analytical statics are inserted here for convenience of reference.

19. Composition and Resolution of any Number of Forces Applied at a Point. (*a*) Forces in one Plane.—Referring to Fig. 1, it will be seen that if AC represents any force R, then AD and AB will represent the rectangular components of this force. Hence calling these components F_1, F_2 respectively and denoting by α the angle made by F_2 with R, we have

$$F_1 = R \sin \alpha, \quad F_2 = R \cos \alpha.$$

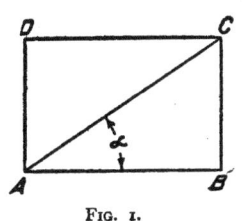

FIG. 1.

It will also appear from a consideration of Fig. 1 that if several forces F_1, F_2, F_3, etc., in the same plane and applied at a point, make angles α_1, α_2, α_3, with the axis of X, the sum of their components resolved parallel to X is $F_x = \Sigma F \cos \alpha$, and the sum of their components parallel to Y is $F_y = \Sigma F \sin \alpha$. Since the resultant $R = \sqrt{F_y^2 + F_x^2}$, we have for its value $R = \sqrt{(\Sigma F \sin \alpha)^2 + (\Sigma F \cos \alpha)^2}$.

Also, if θ is the angle made by R with X, $\tan \theta = \dfrac{\Sigma F \sin \alpha}{\Sigma F \cos \alpha}$.

For equilibrium $\Sigma F \sin \alpha = 0$, $\Sigma F \cos \alpha = 0$.

(*b*) Forces in Space.—For a force in space it is easy to see that if the point of application be taken as the origin of coördinates, the components F_x, F_y, F_z, along the axes are respectively $F_x = F \cos \alpha$, $F_y = F \cos \beta$, $F_z = F \cos \gamma$, α, β, γ being the direction angles of the force F. Also, $F^2 = F_x^2 + F_y^2 + F_z^2$. Hence for a system of forces

$$R = \sqrt{(\Sigma F \cos \alpha)^2 + (\Sigma F \cos \beta)^2 + (\Sigma F \cos \gamma)^2}.$$

Calling α_r, β_r, γ_r, the direction angles of R, we have

$$\cos \alpha_r = \frac{\Sigma F_x}{R} = \frac{\Sigma F \cos \alpha}{R},$$

$$\cos \beta_r = \frac{\Sigma F_y}{R} = \frac{\Sigma F \cos \beta}{R},$$

$$\cos \gamma_r = \frac{\Sigma F_z}{R} = \frac{\Sigma F \cos \gamma}{R}.$$

The conditions of equilibrium with such a system are

$$\Sigma F \cos \alpha = 0, \ \Sigma F \cos \beta = 0, \ \Sigma F \cos \gamma = 0.$$

20. Center of Parallel Forces.—The center of a system of parallel forces is the point through which the resultant always passes, however the direction of the component forces, while still remaining parallel, may be shifted.

It may be found as follows:—

Suppose x_r, y_r, z_r, to be the coördinates of the center, x, y, z, being those of the point of application of any component F. If we assume the component forces to be made parallel to Y, it follows that $x_r \Sigma F = \Sigma Fx$, since the moment of the resultant $R = \Sigma F$, relatively to the axis Z, must be equal to the sum of the moments of the various components relatively to the same axis, *i.e.*, to ΣFx. In like manner, if the components be made parallel to Z and to X, $y_r \Sigma F = \Sigma Fy$, $z_r \Sigma F = \Sigma Fz$. These equations give the values of x_r, y_r, z_r, and so determine the position of the center.

21. Center of Mass, Center of Gravity, Center of Inertia. —These terms all apply to the same point, which is the center of the system of parallel forces formed by the weights of the particles composing the body or system of bodies.

It follows from the last preceding demonstration that for a system of masses the coördinates of the center of mass are

$$x_r = \frac{\Sigma Wx}{\Sigma W} = \frac{\Sigma Mx}{\Sigma M}, \quad y_r = \frac{\Sigma Wy}{\Sigma W} = \frac{\Sigma My}{\Sigma M}, \quad z_r = \frac{\Sigma Wz}{\Sigma W} = \frac{\Sigma Mz}{\Sigma M}.$$

For a single body of mass M,

$$x_r = \frac{\int x dM}{\int dM}, \quad y_r = \frac{\int y dM}{\int dM}, \quad z_r = \frac{\int z dM}{\int dM}.$$

If the body is homogeneous,

$$x_r = \frac{\int x\,dV}{\int dV}, \quad y_r = \frac{\int y\,dV}{\int dV}, \quad z_r = \frac{\int z\,dV}{\int dV}.$$

The integrals must, of course, be taken between the proper limits.

If the origin is at the center of mass, $x_r = 0$, $y_r = 0$, $z_r = 0$; whence

$$\int x\,dM = 0, \quad \int y\,dM = 0, \quad \int z\,dM = 0.$$

PARTICULAR CASES OF RECTILINEAR MOTION.

22. Uniform Motion.—It appears from the foregoing that a body may remain at rest either when acted upon by no force whatever, or under the action of a system of balanced forces, since in each case the effective force acting is ò. The second case is the only one actually realized.

Also a body may move uniformly in a straight line, either under the action of no force or under the action of a system of balanced forces. In the second case, the only one actually realized, the body is often said to be in dynamical equilibrium.

Illustrations of such a condition of motion are found in the uniform motion of a steamer when the motive force is just equal to the resistance of the water and air; and in the case of a railway train when the motive force is just balanced by frictional and other resistances.

23. Uniformly Variable Motion.—A body acted upon by an unbalanced force moves with a variable motion. The most important case is that in which the force is constant and in the direction of the body's motion, in which case the acceleration is constant and the motion is uniformly accelerated.

That the acceleration, $\dfrac{dv}{dt} = a$, is constant follows immediately from the 2d Law of Motion, since the effect of a constant force is to produce a constant rate of change of speed.

It is furthermore true that $a = \dfrac{dv}{dt} = \dfrac{d^2 s}{dt^2}.$ Hence by integration between the limits o and t we reach the familiar formulæ for uniformly accelerated motion when the body starts from a state of rest under the action of the accelerating force, viz.:—

$$v = at \quad (1) \qquad s = \tfrac{1}{2} at^2 \quad (2) \qquad v = \sqrt{2\,as} \quad (3).$$

Equation (3) is obtained by eliminating t from (1) and (2).

For freely falling bodies we have $v = gt, \qquad h = \tfrac{1}{2} gt^2, \qquad v = \sqrt{2gh,}$ where h is the distance fallen through and g the acceleration due to gravity.

If the body possesses an initial velocity v_0 when the accelerating force begins its action, $v_t = v_0 + at$, $s_t = v_0t + \frac{1}{2}at^2$, v_t, s_t, being respectively the velocity acquired and space traversed in t seconds.

If the force acts in opposition to an already-acquired motion, the acceleration is negative and the motion is uniformly retarded. In this case

$$v_t = v_0 - at, \qquad s_t = v_0t - \frac{1}{2}at^2. \text{ Also } s_t = \frac{v_0^2}{2a}.$$

24. Measure of Impulse.—The origin of the equation $Ft = MV$ will now be clear. If a force F acts for a time t on a mass M, it is evident that the resulting change in velocity is $V = at$. But $F = Ma$; whence $Ft = Mat = MV$. That is, the impulse is measured by the resulting change of momentum.

25. Comparison of Forces with Gravity.—If a force F acts on a mass of weight W, the acceleration a imparted can be determined from the proportion $F : W :: a : g$, whence $a = g\dfrac{F}{W}$.

26. Motion over Inclined Plane.—It will be seen, by a reference to Fig. 2, that for a body of weight W descending a frictionless inclined plane, of height H and length L, the accelerating force will be $F = W\dfrac{H}{L}$.

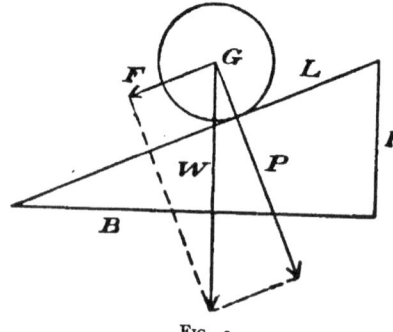

FIG. 2.

Hence, as $F : W :: a : g$, the acceleration of the body will be $a = g\dfrac{H}{L}$.

As a is constant, the motion will be uniformly accelerated. By substituting this value of a in the general formulæ for uniformly accelerated motion, we derive formulæ for the case of a body descending an inclined plane, as follows:—

$$v = gt\frac{H}{L}, \quad s = \tfrac{1}{2}gt^2\frac{H}{L}, \quad v = \sqrt{2gs\frac{H}{L}}.$$

Denoting the angle of slope of the inclined plane by θ, we have $\dfrac{H}{L} = \sin \theta$ and hence

$$v = gt \sin \theta, \quad s = \tfrac{1}{2}gt^2 \sin \theta, \quad v = \sqrt{2gs \sin \theta}.$$

ADDITIONAL PROPOSITIONS.

27. Resultant Momentum.—The resultant momentum in any direction of a system of bodies is $M_r = \Sigma mv$, v being the resolved component of the velocity of m in the direction considered. Thus the resultant momentum of a system of particles parallel to X is $M_x = \Sigma mv_x = \Sigma mv \cos \alpha$, that parallel to Y, $M_y = \Sigma mv_y = \Sigma mv \cos \beta$, that parallel to Z, $\Sigma M_z = \Sigma mv_z = \Sigma mv \cos \gamma$. Also $M_r = \sqrt{M_x{}^2 + M_y{}^2 + M_z{}^2}$.

28. Relation to Center of Mass.—The resultant momentum of any system of bodies is the same as if they were concentrated at the center of mass of the system.

Let m_1, m_2 be the masses of two bodies at distances l_1, l_2 from any assumed coördinate plane, and let l_g be the distance of their center of mass from the same plane. Then, since the moment, relatively to the plane, of the sum of the masses assumed to be concentrated at their center of mass is the same as the sum of their separate moments (see § 21, p. 6), we have $(m_1 + m_2) l_g = m_1 l_1 + m_2 l_2$.

Taking the derivative of each term relatively to time we have $(m_1 + m_2)$

$$\frac{dl_g}{dt} = m_1 \frac{dl_1}{dt} + m_2 \frac{dl_2}{dt}.$$ But the derivatives represent velocities of m_1, m_2,

separately and of $(m_1 + m_2)$ assumed to be concentrated at their center of mass. Calling these v_1, v_2, v_g, respectively we have $(m_1 + m_2) v_g = m_1 v_1 + m_2 v_2$. The second member of the equation represents the resultant momentum of m_1, m_2, and the first member represents the momentum of the masses concentrated as assumed.

The proof is general, for with any number of additional masses whatsoever we may follow the same process, combining them successively. Thus we may apply the same reasoning as before to a third mass m_3 in connection with the masses m_1, m_2 supposed to be concentrated at their center of mass; and so on.

It also follows from what precedes that the resultant momentum of any system of bodies relatively to their center of mass is zero.

29. Couple.—If two equal and opposite parallel forces, not acting in the same straight line, are applied to a body, their algebraic sum is zero, and hence there is no tendency toward motion of translation. But as their resultant moment relative to any point can never become zero, their sole effect will be to cause rotation about an axis.

The rotary effect, torque or moment of a couple is measured by the product of either force into the length of the arm.

Let F, F' (Fig. 3) constitute a couple whose arm is AB. To find the rotary effect, let P be any point. The moments of F, F', relatively to P, are $F \times AP$, $F' \times BP$. Hence, the total resultant moment of the two forces is $F \times AP + F' \times BP = F \times AB$.

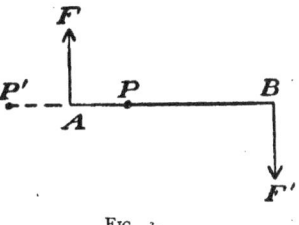

FIG. 3.

If the point be taken on the same side of both forces, as at P', the resultant moment is $F' \times P'B - F \times P'A = F \times AB$, as before. Hence $T = Fl$, where l is the arm.

A couple may be turned in its own plane or moved parallel to itself without altering its efficiency, since the product Fl remains constant. Also, any couple may be replaced by an equivalent one having a given arm. If l_0 be the arm, the corresponding force will be $F_0 = \dfrac{T}{l_0}$.

30. Combination of Couples having the Same Axis.—*The resultant moment of any number of couples lying in the same or parallel planes is equal to the algebraic sum of their separate moments.*

Calling T_r the resultant moment, $T_r = \Sigma Fl$. For equilibrium, so far as rotation is concerned, $\Sigma Fl = 0$.

Couples not having the same axis may be compounded by a process similar to that used in compounding oblique forces. (See § 94, p. 40.)

Evidently a couple cannot be balanced by any single force, but only by the application of an equal and opposite couple.

31. Total Effect of Force on Free Body.—The tendency of any force acting upon a body is, in general, to produce (1) a translatory motion, and (2) a rotation. If the force acts through the center of mass of the body, the resultant moment relatively to that point is zero, and there is no tendency to rotation if the body is free. This will be clear if we consider the force in question to be resolved into an infinite number of parallel forces applied to each particle of the body, and proportional to the mass of that particle. These would tend to cause equal accelerations in each particle, and hence to cause all points in the body to move with the same velocity.

Because of this property of the center of mass it is often called the *center of inertia*.

If the line of action of the force does not pass through the center of mass, the total translatory effect will be the same as if the force were applied at that point, and the rotary effect will be equal to its moment with regard to the center of mass. In Fig. 4 let F be a force applied at A. The condition of the body will not be altered if we imagine two opposite forces, F_1, F_2, each equal and parallel to F, to be applied at G, the center of gravity. But we have now

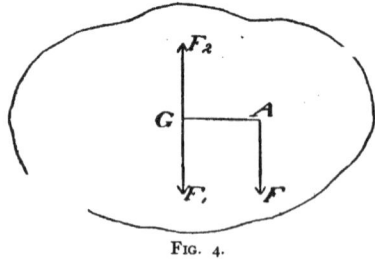

FIG. 4.

a couple whose moment is $F \times AG$, and a force F_1 equal and parallel to F, and passing through G.

. It can be shown that the rotation produced by the couple $F \times AG$ will take place about an axis passing through the center of mass of the body.

Every possible motion of a rigid body may be considered as compounded of a rotation about its center of mass combined with a translatory motion.

It follows from what precedes that any number of forces acting upon a body or system of bodies may be replaced by a single resultant force passing through the center of mass, and a resultant couple.

32. Transferability of Force.—Since the translatory effect of a force is solely to produce motion in its line of action while the rotary effect is its moment, the total mechanical effect of a force will not be altered by transferring its point of application to any point in its line of action.

This principle facilitates the process of composition of two or more oblique forces not applied at the same point. The lines of action of any two components may be prolonged till they meet in a point, in which case the Parallelogram of Forces becomes directly applicable. The remaining components can be combined successively in the same manner.

MEASUREMENT OF FORCE AND MASS.

33. Units and Standards.—As a preliminary to the discussion of this subject it will be necessary to explain briefly the nature of the units of length, mass, and time which are adopted in scientific and industrial measurements.

The unit of length in the metric system is the distance between two transverse lines ruled on the surface of a certain standard platinum-iridium bar (a line standard) adopted as such in 1889 and kept at Paris, which is known as the International prototype standard meter. The standard temperature at which the bar is correct is $0°$ C. This length is as exactly as possible the length at the same temperature of the original "Mètre des Archives" of 1799, which is an end standard. For scientific purposes, however, the centimeter, one one-hundredth part of the meter, has been found to be preferable to the meter for use as a unit.

- The unit of length in the British system is the Imperial yard, which is the distance between two transverse lines ruled on gold studs inserted in a certain bronze bar (a line standard), when at a temperature of $62°$ F. This bar was made the legal standard of Great Britain in 1855 and is kept in the Standards Office, Westminster, London. The foot is habitually employed as a more convenient practical unit than the yard.

The legal yard of the United States is defined (1893) as being of such length that 39.37 of its inches shall equal the length of the meter.

The unit of mass in the metric system is the gram. This is a mass equal

to one one-thousandth part of the mass of the International standard prototype kilogram, a platinum-iridium cylinder, adopted as a standard in
1889, and placed together with the International prototype meter in the
care of the International Metric Commission at St. Cloud, Paris.   This kilogram is an exact copy of the original "Kilogramme des Archives" of 1799.

The British unit of mass is the Imperial avoirdupois pound.   This is
represented by a certain cylindrical piece of platinum, legalized in 1855
and kept at Westminster.

The legal avoirdupois pound of the United States is a pound of such
mass that 2.2046 pounds shall equal one kilogram.

Copies of the prototype standard meter and kilogram ("national prototype standards") are in the possession of each of the principal nations.
Those belonging to the United States are kept at the Office of Weights
and Measures in Washington.

The standard meter is found by comparison to be equal to 39.370113
inches of the British standard yard.   The standard kilogram is equal
to 15432.3564 grains of the British standard pound.   This is the legally
adopted ratio in Great Britain.

The unit of time universally employed in both the Metric and British systems is the ordinary sexagesimal second of civil life, which is the $\frac{1}{86400}$ part
of the length of the mean solar day.

It was originally intended that the meter should be one ten-millionth
part of a quadrant of the meridian in length.   The actual meter, however,
is slightly shorter than this.   There is, moreover, apparently a small
difference in the length of the different meridians according to the longitude.
The mean length of a quadrant of the meridian is in fact 10,002,000 meters
instead of 10,000,000 meters.   (Clarke, 1880.)

It was also intended that the gram should be precisely the mass of a
cubic centimeter of water at the temperature of its maximum density,
4° Centigrade.   The actual mass of a cubic centimeter of water at 4° C.
is, in fact, 0.999972 grams as determined at the International Metric
Bureau (1907).

Inasmuch as there is always a possibility of a slow secular change in
the molecular constitution of a metal bar and consequently of a minute change
in its length, the use of the wave-length of a selected kind of homogeneous
light as a standard has been suggested at different times by several persons,
first by Lamont (1823).   Michelson has recently (1892-93) determined
with great accuracy and by a direct method the length of such a wave,
produced by incandescent cadmium vapor, in terms of the present prototype standard meter, and hence determined the length of the meter in
terms of this wave-length.

The meter contains 1553163.6 wave-lengths of a certain component
(red) ray of cadmium light.

By repeating this measurement many years hence it will be possible to
ascertain whether the standard meter has in any way altered in length.

Also if the present standard and all accurate copies of it were to be destroyed it could still be reproduced at any time from this wave-length.

In view of the possible question as to a sensible variation in the length of the day, other units of time than the second have been proposed, as, for example, the period of vibration of an elastic spring of determined material and dimensions under fixed conditions of temperature, etc., "a perennial spring," to use Kelvin's term. A tuning-fork would be the most suitable instrument for the purpose. But aside from uncertainty as to exact constancy of external conditions, the liability to a secular change of rate of the fork arising from slow internal molecular changes would make the use of such a standard unpractical.

It was suggested by Maxwell that the period of vibration of a homogeneous light-wave emitted by incandescent vapor, that of sodium, for instance, might be used. This period is exact and has been accurately determined in terms of the present second. An objection, however, is its excessive smallness, approximately 510×10^{-12} seconds.

34. Dynamical Measure of Force.—Any mechanical force, whether pressure or impulse, may be measured by means of a system based upon the 2d Law of Motion.

By making a suitable choice of units we may write $F = Ma$ instead of merely $F \propto Ma$.

The unit of force chosen is that force which produces an acceleration unity in a mass unity. This is indicated algebraically in the equation. $F = Ma$, as by making M and a each equal to unity, F also becomes equal to unity.

Since $F = Ma$, it follows also that $M = \dfrac{F}{a}$, that is, the mass of a body may be measured by the constant ratio between any force and the acceleration which that force will produce when acting on the body.

Hence, if W is the force by which a given body of mass M is drawn toward the earth, that is, if W is the weight of the body, and g the acceleration due to gravity, we have, since g is an acceleration produced by a force W acting upon that body, $M = \dfrac{W}{g}$.

The unit of mass in all systems will weigh (i. e., be drawn to the earth by a force of) g units of force. For, since $M = \dfrac{W}{g}$, then, when M equals unity, W must be numerically equal to g; that is, the unit of mass must be drawn to the earth by a force of g units of force. Likewise, the unit of force must be $\dfrac{1}{g}$ part of the weight of the unit of mass.

It is evident that the equation $M = \dfrac{W}{g}$ is independent of the particular units of mass, force, and length that we may choose to adopt.

It will be clear from what has been said that either the unit of mass or the unit of force may be selected arbitrarily, and the corresponding unit of force or of mass determined in accordance with the relations which have been shown to exist between them.

Two principal systems of measurement of force and mass have been used, called respectively the *Absolute* and the *Gravitation* System. The former selects the unit of mass and determines the corresponding unit of force; the latter selects the unit of force and determines the corresponding unit of mass.

35. Absolute System.—In this system, the *mass* of the gram or pound is chosen as the unit of mass. The mass of a body is therefore expressed either in grams or in pounds.

The unit of force is determined from this, it being $\dfrac{1}{g}$ part of the *weight* of the gram or pound. The value of g is commonly expressed in centimeters or in feet, according as metric or British measures are used.

These units were originally proposed by Gauss. It is evident that they are absolutely constant in all places and under all conditions.

The Absolute System is used in all refined physical measurements.

We shall see later that units of measurement of all physical forces can be derived from the fundamental units of length, mass, and time. Such units are called *absolute units*, or, more logically, *derived units*.

The particular form of absolute system now universally employed in science is the *Centimeter-gram-second System* (*C. G. S. System*), so called from the fundamental units of length, mass, and time on which it is based.

The *C. G. S.* unit of force is called a *dyne*, and is $\dfrac{1}{g}$ of the weight of a gram, g being expressed in centimeters. The British *Foot-pound-second System* is based upon the foot, pound, and second. The corresponding unit of force is called a *poundal*, and is $\dfrac{1}{g}$ of the weight of a pound, g being expressed in feet. This system, however, is practically obsolete.

Various other forms of the Absolute System have been used in past times; as, for example, a *meter-gram-second system*, a *millimeter-gram-second system*, and a *foot-grain-second* system.

It will appear on consideration that the absolute unit of force thus derived will fulfil the requirements of the general definition of a unit of force, and generate an acceleration unity in a mass unity. Thus, for example, the dyne must generate an acceleration of one centimeter in a mass of one gram. For if such a mass falls freely, it will acquire an acceleration of g centimeters. But the accelerating force is g dynes, that being the weight of a gram. Hence, under the action of a force of one dyne the acceleration acquired would be one centimeter.

. As g has a value of 980.9 cm. at Paris and the absolute unit of force is equal to $\dfrac{1}{g}$ of the weight of the unit of mass *at any place whatever*, the dyne is equal to $\dfrac{1}{980.9}$ of the weight (force of attraction to the earth) of a gram *at Paris*. And as the value of g at London is 32.191 ft., the poundal is equal to $\dfrac{1}{32.191}$ of the weight of a pound *at London*.

36. Gravitation System.—In this system the *weight* of the pound or gram (that is, the force by which the pound-weight or gram-weight is drawn toward the earth) is chosen as the unit of force. The weight of the kilogram is also frequently used as a unit.

The mass of a body is therefore measured by its weight *in pounds* or *grams* or *kilograms*, divided by the acceleration produced by gravity. This quotient is evidently a constant for any particular body; for, if its weight W (that is, the force by which it is attracted to the earth) varies from any cause (as from change of latitude, change of altitude, etc.), the acceleration g will vary proportionally. (Law II.)

Since in all systems the unit of mass weighs g units of force, it follows that the unit of mass in the Gravitation System must weigh g pounds, grams, or kilograms, according to which of these is employed as a unit of force. That is, its mass must be g times that of the pound-weight, gram-weight, or kilogram-weight.

In the Gravitation System g is usually expressed in feet or in meters.

The unit of mass in the Gravitation System has never acquired any distinct name, although the term *Slug* has been suggested. When this system is used, care must be taken to express the mass of the body in terms of this unit and not in pounds or grams.

We shall see hereafter that the weight of the pound or gram, and consequently g, vary with the locality, so that the units of both force and mass in the Gravitation System are variable unless the definite value of g at some particular place is assumed as a standard, which has not been customary. This want of definiteness constitutes a fatal objection to the use of the Gravitation System where the greatest accuracy is required. It is commonly used, however, in ordinary engineering computations in which the variations of g may be neglected.

The value for g of 32.2 feet or 9.8 meters may always be employed in the Gravitation System, no closer approximation being necessary.

The expression for a force in gravitation units may be transformed into absolute units by multiplying the numerical value of the force in gravitation units by the value of g for the locality at which the measurements are made. Methods of determining the value of g for any place will be explained later.

The student will notice that in the discussion of this subject ambiguity is likely to arise from the fact that the terms "pound" and "gram" are used in two different senses, both as denoting a mass, and as denoting the force by which this mass is drawn toward the earth. To remove this ambiguity as far as possible, the term "weight of a pound or gram" has been used in the preceding discussion whenever a force is referred to. The pound and gram are primarily standards of mass, and the use of the same names to denote forces is a secondary application. Whenever the terms "pound" or "gram" are used as denoting forces, they are to be understood as really meaning the *weight* of the units of mass denoted by those names.

In the equation $W = Mg$, which is true for all systems of measurement of mass and force, W expresses in units of force the downward tendency (due to its weight) of a body of mass M. W and M must always be expressed in units of the same system, either Gravitation or Absolute. Also in both systems g expresses the number of units of force by which a unit mass is drawn to the earth.

37. Example.—The application of the absolute system may be made clearer by a simple example.

Let us suppose that a certain force A as measured in Paris is equal to the weight of 100 grams at that place. A certain other force B as measured at London is equal to the weight of 1,000 grams at that place. Since the force by which a gram is attracted to the earth is not the same at these different localities, we cannot compare the forces directly. If, however, we reduce the forces to absolute units, the difficulty is avoided. The force A expressed in absolute *C. G. S.* units is $A = 100 \times 980.9 = 98,090$ dynes, and also $B = 1,000 \times 981.2 = 981,200$ dynes; since the value of g for Paris is 980.9 cm., and that for London is 981.2 cm. These results immediately show the absolute relation of the forces A and B.

It is evident that an absolute system may be based upon any convenient units of length, mass, and time. Also a system of measurement might be used in which the unit of force was taken as the invariable weight of the standard pound or gram at some particular place.

A table of the units most commonly used will be found on p. 19.

38. Note.—A philosophical study of the nature of force and of our knowledge of it, shows that ultimately we have no way of measuring the relative mechanical magnitude of all kinds of forces except by their effect in generating momentum. Hence, Law II. of Motion is rather a definition of a system of measurement than a law determined from measurements obtained by other means. We call forces mechanically equal when they produce an equal rate of change of momentum; and if one force produces a rate of change n times as great as that produced by a second force, we call the former force n times as great as the latter.

TABLE OF DIMENSIONS.

Quantities	Dimensions
Length	L
Mass	M
Time	T
Area	L^2
Volume	L^3
Angle $= \dfrac{\text{Arc}}{\text{Radius}}$	L°
Velocity $= \dfrac{S}{T}$	LT^{-1}
Angular Velocity $= \dfrac{\theta}{T}$	T^{-1}
Acceleration $= \dfrac{V}{T}$	LT^{-2}
Momentum $= MV$	MLT^{-1}
Force $= Ma$	MLT^{-2}
Torque $= Fl$	ML^2T^{-2}
Energy $= \frac{1}{2}MV^2$	ML^2T^{-2}
Power $= \frac{1}{2}\dfrac{MV^2}{T}$	ML^2T^{-3}
Moment of Inertia $= Mk^2$	ML^2

ENERGY.

39. Work and Energy.—Work is performed whenever a force produces motion in opposition to a resistance; or more generally whenever a force acts to move its point of application through space.

Work is found to be directly proportional to each of these variables, force and distance. Hence representing the force by F and the distance through which its point of application is moved by S, the work done is expressed by the equation $E = FS$.

"Energy is the capacity of doing work."—MAXWELL.

40. Units.—The unit of work in practical use by engineers is the *kilogrammeter* in Metric, and the *foot-pound* in British measures. The kilogrammeter is the work done in overcoming a force equal to the weight of one kilogram through the space of one meter. The foot-pound is the work done in overcoming a force equal to the weight of one pound through the space of one foot. The kilogrammeter is equal to 7.233 foot-pounds. The *gram-centimeter* is frequently employed as a smaller unit of work. The *foot-ton* is often used when great amounts of energy are to be considered.

In purely scientific investigations, absolute units are commonly employed. The absolute unit of work is the work done in overcoming an absolute unit of force through a unit of length.

The unit of work in the centimeter-gram-second system is called the *erg*, and is the work done in overcoming a force of one dyne through one centimeter. It is a *dyne-centimeter*. In the foot-pound-second system, the unit is the *foot-poundal*, which is the work done in overcoming a force of one poundal through one foot.

The rate of work or work-rate is the work done in a unit of time. That is, $R = \dfrac{FS}{T}$. If the rate is not uniform, this definition applies to its average value. In general, $R = \dfrac{dE}{dt}$.

Lord Kelvin has given the title "activity" to the rate of doing work. It is usually called "power" by engineers.

A standard in ordinary practical use, for comparing the rate of work or the activity of different motors, is the *horse-power*, which is equal to 33,000 foot-pounds per minute, or 550 foot pounds (76.0 kilogrammeters) per second. One horse-power is equal to 7.46×10^9, or 7,460 million ergs per second, assuming $g = 981$ cm. The French *force de cheval* or *cheval-vapeur*, is not identical with the English *horse-power*, but is defined as 75 kilogrammeters per second. This equals 7.36×10^9 ergs per second, if $g = 981$ cm.

A unit of work frequently used in connection with electrical measurements is the *Joule* = 10,000,000 (10^7) ergs. The corresponding unit of activity is the *Watt* = 10^7 ergs per second. The *kilowatt*, a unit commonly used in rating electric machinery, is 1,000 watts = 1.34 horse-power.

Engineers often use the *horse-power-hour*, the *watt-hour*, the *kilowatt-hour* respectively as practical units of work. A horse-power-hour is the work done in one hour when energy is expended at a constant rate of one horse-power. The other units referred to are defined in like manner.

41. Accumulated Work. Kinetic Energy.—Frequently the space S in the formula for work done is not directly known, while the velocity with which a body is moving is given. An expression for the work which can be done by a body of mass M, moving with a velocity V, may be found by ascertaining the space through which the body would move against a constant resistance and with a uniformly retarded motion before its velocity would be reduced to zero, and substituting this space as expressed in terms of V in the general equation.

If the body is moving with a velocity V against a resistance F which is capable of producing a retardation a, the distance over which the body will move is $S = \dfrac{V^2}{2a}$. Substituting this value, we have $E = FS = F\dfrac{V^2}{2a}$.

But $\qquad\qquad \dfrac{F}{a} = \dfrac{W}{g}$, whence $E = \dfrac{W}{g}\dfrac{V^2}{2} = \tfrac{1}{2}MV^2$.

Hence *the work which a moving body is capable of performing in virtue of its motion is equal to half its mass into the square of its velocity.*

The result is expressed in units of either the absolute or the gravitation system, according as the mass is expressed in the one or the other of these.

It will be observed that in the preceding demonstration it is assumed that the work necessary to bring a body to rest is the same, whatever may be the nature of the resistance overcome, which is an experimentally-determined fact.

The energy possessed by a body because of its motion is called Kinetic Energy.

The product MV^2, which is twice the kinetic energy of a moving mass, was called by Leibnitz the *vis viva* of the mass.

TABLE OF UNITS.

System	Mass	Length	Time	Force	Energy or Work	Power or Activity
Absolute Metric C. G. S.	Gram	Centimeter	Second	Dyne	Erg = Dyne-centimeter	$\dfrac{\text{Erg}}{\text{Second}}$
					Joule = 10^7 Ergs.	Watt = $\dfrac{\text{Joule}}{\text{Second}}$ = Volt-ampere = $10^7 \dfrac{\text{Erg}}{\text{Second}}$
					Watt-hour Kilowatt-hour	Kilowatt = 1,000 Watts
Gravitation Metric	"Metric Slug" weighing 9.8 kg.	Meter	Second	Kilogram.	Kilogram-meter.	Force de Cheval $= 75 \dfrac{\text{Kilogrammeter}}{\text{Second}}$
Gravitation British	"Slug" weighing 32.2 lb.	Foot	Second	Pound	Foot-pound Horse-power-hour	Horse-power = $550 \dfrac{\text{Foot-pound}}{\text{Second}}$

42. Rotating Bodies.—It remains to be shown how the kinetic energy of a rigid rotating body is determined. The *angular velocity* of such a body is measured by the angle in circular measure through which it rotates in a unit of time, and is numerically equal to the linear velocity of any point at a distance unity from the axis of rotation. That is, the angular velocity $\omega = \dfrac{\theta}{t}$ or in general $\omega = \dfrac{d\theta}{dt}$, where θ is the angle described in t seconds.

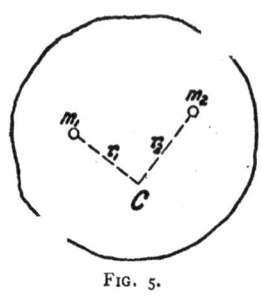

FIG. 5.

Let m_1 be the mass of a particle of the body situated at a distance r_1 from the axis of rotation through C (Fig. 5), and let ω be the angular velocity of the body. The energy of the particle m_1 is $\frac{1}{2}m_1v_1^2$, or $\frac{1}{2}m_1r_1^2\omega^2$, since $v_1 = r_1\omega$.

The energy of any other particle of mass m_2 at a distance r_2 from C is $\frac{1}{2}m_2r_2^2\omega^2$. Similar expressions may be obtained for all the particles of the body. Hence the total energy of rotation of the body is

$$E = \tfrac{1}{2}m_1r_1^2\omega^2 + \tfrac{1}{2}m_2r_2^2\omega^2 + \tfrac{1}{2}m_3r_3^2\omega^2 \ldots + \tfrac{1}{2}m_nr_n^2\omega^2 = \tfrac{1}{2}\omega^2\Sigma mr^2;$$

whence the formula $E = \frac{1}{2}\omega^2\Sigma mr^2$ represents the accumulated energy of the body due to its rotation about C, and is the amount of work which it will perform while being brought to rest.

43. Moment of Inertia.—The expression Σmr^2 is of very frequent occurrence in dynamics, and is known as the *moment of inertia* of the mass. Denoting it by I, we have $E = \frac{1}{2}\omega^2 I$. That is, *the energy of rotation of a body is equal to its moment of inertia multiplied by half the square of its angular velocity.* The quantity I evidently varies with the form of the body, its mass, and the position of its axis of rotation.

Using the notation of the Calculus we may write $I = \int r^2 dm$.

44. Radius of Gyration.—If we assume the total mass of a body to be concentrated in a single point situated at a distance k from the axis of rotation such that the moment of inertia of the mass thus concentrated shall equal the moment of inertia of the distributed mass of the body, then

$$Mk^2 = I \text{ and } k = \sqrt{\dfrac{I}{M}}.$$

The radius k is called the *radius of gyration* of the mass.

45. Moment of Inertia about any Axis.— Knowing the moment of inertia I_g with reference to an axis passing through the center of gravity G of a mass it is often necessary to find the moment of inertia I_a with reference to an axis parallel to the former and passing through a point A at a given distance d from G.

In Fig. 6 let m be any element of mass, i_g its moment of inertia relatively

to an axis G, i_a its moment of inertia relatively to an axis through A, and denote by r_g and r_a the respective distances of m from G and A.
Let $GC = x$, $mC = y$. Then
$i_g = mr_g^2 = m\,(x^2 + y^2)$. Also
$i_a = mr_a^2 = m\,[(x + d)^2 + y^2]$
whence $i_a - i_g = m\,(2dx + d^2)$
$= 2\,d\,m\,x + m\,d^2$. For any
other element m', $i'_a - i'_g =$
$(2\,d\,m'\,x' + m'd^2)$, and for the
whole mass $\Sigma i_a - \Sigma i_g = 2\,d\,\Sigma\,m\,x$

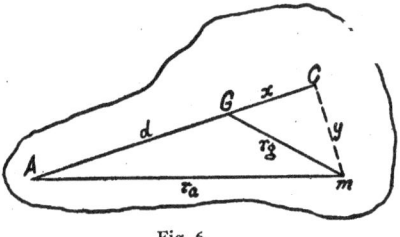

Fig. 6.

$+\ d^2\Sigma m$. But $\Sigma\,m\,x = 0$, since this is the sum of the values of mx for all the particles composing the body with reference to an axis through the center of gravity. Hence $I_a - I_g = Md^2$ and $I_a = I_g + Md^2$.

46. Determination of Angular Velocity.—The angular velocity generated by the expenditure of an amount of work FS upon a body can be found as follows:—

$$FS = \tfrac{1}{2}\omega^2 I;\ \text{whence}\ \omega = \sqrt{\frac{2FS}{I}}\,.$$

It will also be seen that by solving this equation relatively to I an expression may be found for the moment of inertia, thus indicating a way in which the value of this can be determined experimentally.

TABLE OF MOMENTS OF INERTIA.

FROM ROUTH'S RIGID DYNAMICS.

The moment of inertia of
(1) A rectangle whose sides are $2a$ and $2b$,
    about an axis through its center in its plane
    perpendicular to the side $2a = \text{mass} \times \dfrac{a^2}{3}$,
    about an axis through its center
    perpendicular to its plane $= \text{mass} \times \dfrac{a^2 + b^2}{3}$.
(2) An ellipse, semi-axes a and b,
    about the major axis $a = \text{mass} \times \dfrac{b^2}{4}$,
    about the minor axis $b = \text{mass} \times \dfrac{a^2}{4}$,
    about an axis perpendicular to its plane
    through the center $= \text{mass} \times \dfrac{a^2 + b^2}{4}$.

In the particular case of a circle of radius a, the moment of inertia about a diameter $= \text{mass} \times \dfrac{a^2}{4}$, and that about a perpendicular to its plane through the center $= \text{mass} \times \dfrac{a^2}{2}$.

(3) An ellipsoid, semi-axes a, b, c,

about the axis $a = \text{mass} \times \dfrac{b^2 + c^2}{5}$.

In the particular case of a sphere of radius a the moment of inertia about a diameter $= \text{mass} \times \dfrac{2}{5} a^2$.

(4) A right solid whose sides are $2a$, $2b$, $2c$,

about an axis through its center perpendicular to the plane containing the sides b and c $\Big\} = \text{mass} \times \dfrac{b^2 + c^2}{3}$.

These results may all be included in one rule, as an assistance to the memory.

Moment of inertia about an axis of symmetry $\Big\} = \text{mass} \times \dfrac{\text{(sum of squares of perpendicular semi-axes)}}{3, 4 \text{ or } 5}$.

The denominator is to be 3, 4, or 5 according as the body is rectangular, elliptical or ellipsoidal.

47. Kinetic Energy.—Energy is either *Kinetic* or *Potential*. Kinetic energy, as already explained (see § 41, p. 18), is the capacity possessed by a body of doing work in virtue of its motion, and is expressed by the equation

$$K = \tfrac{1}{2}MV^2.$$

ILLUSTRATIONS: Projectile in motion; fly-wheel; stream of flowing water actuating undershot water wheel; current of air moving wind-mill.

The term "kinetic energy" is due to Kelvin. It is sometimes called "actual energy," as proposed by Rankine.

48. Effect of Unbalanced Force on Mass.—If an unbalanced force acts to accelerate the motion of a mass, its effect is to add to the kinetic energy of the mass. The increment of kinetic energy, that is, the work stored up, in a given time, is

$$K = \tfrac{1}{2}M(V_1^2 - V_0^2).$$

If the mass, being in motion, moves against a resistance, it does work which is measured by the decrease of kinetic energy; that is,

$$K = \tfrac{1}{2}M(V_0^2 - V_1^2).$$

The increase of kinetic energy produced by the continued action of a force on a mass is equivalent to the force multiplied by the space through which it acts. This will appear from the following demonstration:—

Consider the action of the constant force F on a mass M through a space S. Then as $F = Ma$ and $S = V_0 t + \tfrac{1}{2}at^2$, we have $FS = MV_0 at + \tfrac{1}{2}Ma^2t^2$. This is equal to the gain of kinetic energy in passing through S. For calling K_0 and K_1 the kinetic energy at the beginning and end of S respectively $K_0 = \tfrac{1}{2}MV_0^2$, $K_1 = \tfrac{1}{2}M(V_0 + at)^2$ whence the gain in kinetic energy, $K_1 - K_0 = MV_0 at + \tfrac{1}{2}Ma^2t^2$ which has already been shown to be the value of FS. Hence $FS = \tfrac{1}{2}M(V_1^2 - V_0^2)$.

If the force F is not constant during the time t, we may divide t into a very great number of parts so small that during each one of them the force acting may be considered as constant. Then, for each one of the elementary spaces traversed in an element of time, the proposition will hold. Hence we have for their sum in all cases

$$\int_{S_o}^{S_{\mathrm{I}}} Fds = \tfrac{1}{2} M(V_{\mathrm{I}}^2 - V_o^2.)$$

This equation may be derived directly by the methods of the Calculus as follows:—

$$F\, d S = F\, vdt = F\, at\, dt = (Ma)\, at\, dt. \quad \text{Hence} \int_{S_o}^{S_{\mathrm{I}}} F\, d S = \int_{t_o}^{t_{\mathrm{I}}} Ma^2 tdt =$$

$$\left[\tfrac{1}{2} Ma^2 t^2\right]_{t_o}^{t_{\mathrm{I}}} = \left[\tfrac{1}{2} MV^2\right]_{V_o}^{V_{\mathrm{I}}}. \quad \text{Hence } F(S_{\mathrm{I}} - S_o) = \tfrac{1}{2} M(V_{\mathrm{I}}^2 - V_o^2).$$

The demonstration evidently applies whether the force F is constant or variable.

From a similar mode of reasoning, it follows that when a moving mass M overcomes a resistance F through a space S, the work done by the body results in a loss of kinetic energy as expressed by the equation

$$FS = \tfrac{1}{2} M(V_o^2 - V_{\mathrm{I}}^2).$$

In the first case considered FS is commonly spoken of as the *work done upon the body* by the force F in time T. In the second case *the work is done by the body* against the force F.

49. Total Kinetic Energy.—The total kinetic mechanical energy of a body is the sum of the energy due to translation and that due to rotation; *i. e.*,

$$K_t = \tfrac{1}{2} MV^2 + \tfrac{1}{2}\omega^2 I.$$

For a system of bodies the sum of the energies of each part must be taken.

50. Equivalent Mass.—It will be seen that if a force acts on a mass to produce simultaneous translation and rotation, as, for example, when a ball is caused to roll on a surface, the relation between the work done upon the body and the total resulting kinetic energy is represented by the equation $FS = \tfrac{1}{2} MV^2 + \tfrac{1}{2}\omega^2 I$.

If we denote by M_e a mass such that with the linear velocity V alone it would possess an amount of energy equal to the total energy of M, we have $\tfrac{1}{2} M_e V^2 = \tfrac{1}{2} MV^2 + \tfrac{1}{2}\omega^2 I$. As $V = \omega R$, $M_e = M + \dfrac{I}{R^2}$.

Thus a sphere descending an inclined plane moves with a linear acceleration only $\tfrac{5}{7}$ of that which it would acquire did it not rotate, since for it $M_e = \tfrac{7}{5}M$.

51. Potential Energy.—This is the capacity of doing work possessed by the bodies or particles composing a system in virtue of their relative position. If bodies are so situated that they are acted upon by a force which will produce motion in them on the removal of some restraining force, and thence generate kinetic energy, the system is said to have potential energy. Thus a mass suspended at an elevation will fall as soon as the cord sustaining it is cut. The potential energy in this case is $P = FS$, where F is the weight of the raised mass, and S is its elevation; or $P = \frac{1}{2} MV^2$ where V is the velocity corresponding to a fall from the height S.

ADDITIONAL EXAMPLES: Stretched spring; head of water.

In general the potential energy of a system is measured by the work expended in reducing its energy to zero; or conversely, by the work which must be done to endow the system with the amount of potential energy which it possesses.

Potential energy is sometimes called "energy of configuration or strain." The term "potential energy" was originally suggested by Rankine.

52. Transformation of Potential Energy into Kinetic, and the Reverse.

EXAMPLES: Pendulum; tuning-fork in vibration; hydraulic ram.

53. Conservation of Energy.—*The sum of the kinetic and potential energies of a system of bodies not acted upon by any external force is a constant.* Or, in other words, *the total energy of a system of bodies is not affected by their mutual actions.*

This fact is learned from observation and experiment, and is found to be true without exception.

As an illustration consider the case of a body falling from a height h_0 and which has descended to a height h_1. The potential energy is Wh_1 and the kinetic energy is $\frac{1}{2}MV_1^2 = W(h_0 - h_1)$. The sum of these is evidently a constant and equal to the original potential energy Wh_0, or to the final kinetic energy on reaching the earth.

The same reasoning can be applied to the case of a body thrown upward, a pendulum, a vibrating tuning-fork or string, etc.

54. Conservative System.—"When the nature of a material system is such that if, after the system has undergone any series of changes, it is brought back in any manner to its original state, the whole work done by external agents on the system is equal to the whole work done by the system in overcoming external force, the system is called a *conservative system.*" —MAXWELL.

55. Different Forms of Energy.—There are various forms of energy other than mechanical energy, the only kind which we have thus far considered. Several schemes of classification of these have been suggested. The following is a modification of one given long since by Professor Balfour Stewart:—

I. VISIBLE MECHANICAL ENERGY.

    a. Visible kinetic energy.

    b. Potential energy of visible arrangement.

II. INVISIBLE MOLECULAR ENERGY.

    a. Kinetic energy of absorbed heat.

    b. Radiant energy (kinetic).

    c. Potential energy due to molecular separation by heat.

    d. Potential energy due to chemical separation.

III. ENERGY OF ELECTRICITY.

    a. Potential energy due to electrical charge.

    b. Kinetic energy due to electrical current.

Another mode of classification possessing certain advantages is the following, given by Professor A. A. Noyes:—

1. Kinetic Energy.	5. Electrical Energy.
2. Gravitation Energy.	6. Magnetic Energy.
3. Cohesion Energy.	7. Chemical Energy.
4. Disgregation Energy.	8. Heat Energy.

56. Note.—It is possible that all energy is in fact kinetic, that potential energy, so-called, is always due to undetected motion of some kind, though of an unknown nature. While this has not as yet been proved to be the case, it is a plausible hypothesis.

For example, Kelvin has shown that the phenomena of elasticity and hence potential energy of strain may be accounted for by an assumed rotation of elementary portions of a mass. Again, the explanation of gravitation by Le Sage assumes the bombardment of masses by "ultra-mundane corpuscles."

A familiar example of a case in which potential energy possessed by a mass is in fact due to motion is found in the case of a gas, which is capable of doing work in virtue of its expansive tendency or pressure. But, according to the Kinetic Theory, this pressure is caused by the rapid motion of the molecules of the gas. The potential energy of the gas is really the kinetic energy of its molecules.

57. Energy Transformations.—*Any one of the various forms of energy may give rise, either directly or indirectly, to any other form.*

ILLUSTRATIONS: Heat produced by percussion or friction; steam-engine: electrification by friction; attraction and repulsion of charged bodies: dynamo-electric generator; electric motor: thermo-electricity; wire heated by electricity: heat of chemical combination; dissociation by heat: voltaic battery; electro-chemical decomposition.

The above proposition relates only to the qualitative aspect of the different transformations of energy. It was formerly known as the principle of the "Correlation of Forces," a term introduced by Grove in 1843, at which early date the distinction between force and energy was not clearly recognized.

58. Conservation of Energy.—The following law, which is a statement of the doctrine of the conservation of energy, holds for all known forms of physical energy. "*The total energy of any body or system of bodies is a quantity which can neither be increased nor diminished by any mutual action of these bodies, though it may be transformed into any of the forms of which energy is susceptible.*"—MAXWELL.

59. Illustrations of Principle.—For example, suppose a dynamo-machine to generate a current of electricity which in its turn drives an electric motor. If the work done in driving the dynamo is measured, and also the heat generated mechanically by friction of machinery, resistance of air to moving parts, etc., the total heat generated by the current in the complete circuit, and the total mechanical work done by the motor, then the mechanical work done in driving the generator will equal the total amount of energy developed in the several resulting operations.

Again, if a definite weight of zinc in a battery cell is oxidized and the chemical energy disappearing in the cell is wholly transformed into the electrical energy of the current produced, and this current expends its energy in heating a resistance coil, it will be found that the same amount of heat is liberated in the circuit that would have resulted from the direct oxidation of the same weight of zinc in a calorimeter.

60. Case of Heat.—When a change of state occurs on raising the temperature of a body, there is always an absorption or liberation of heat, corresponding to the expenditure of work (positive or negative) accompanying the change. Thus, when a liquid is caused to assume a vaporous condition, a great amount of heat disappears (*latent heat of vaporization*). This is due to the fact that mechanical work is done by the expenditure of heat, first in separating the particles of the liquid from one another in the act of evaporation or of boiling (*internal work*), and second, in overcoming the atmospheric pressure which resists the expansion which accompanies the change of state (*external work*). When ice is melted, the large amount of heat that disappears (*latent heat of fusion*) is chiefly expended in doing internal work. Increase of either the internal or external work tends to increase the latent heat. Increased external pressure raises the fusing-point of substances which expand on melting because of the resulting increase in the external work which must be performed. An opposite effect occurs with substances which like ice contract on melting. Anything which increases the internal work accompanying a change of state, as, for example, in vaporization, raises the boiling-point. Thus we may suppose that the rise of the boiling-point of liquids when solids are dissolved in them is due to the increased difficulty of separating the liquid from the dissolved solid.

In ordinary evaporation the heat which necessarily disappears on account of the change of state is withdrawn from neighboring bodies, thus lowering the temperature to a greater extent according as the evaporation is more rapid.

61. Mechanical Equivalent of Heat.—The quantity of mechanical work which is capable of generating one unit of heat, is called the *mechanical equivalent of heat*, or *Joule's equivalent*, from the physicist who first determined its value.

The thermal unit commonly employed by engineers in Great Britain and the United States is the amount of heat necessary to raise one pound of water through $1°$ Fahrenheit. Owing to the variation of the specific heat of water with temperature the temperature of the water must be stated. The C. G. S. unit of heat is the *gram-degree*, the temperature being measured in Centigrade degrees. The readings are ordinarily reduced to the standard hydrogen thermometer.

Several determinations later than that of Joule have been made by different persons using various methods. The value obtained by Rowland at Baltimore in 1879 with certain corrections subsequently applied, for the temperature $15°$ C. is J = 778 foot-pounds at Greenwich = 4.19 × 10^7 (41,900,000) ergs.

62. Energy Factors.—It is found that with all forms of energy two factors, upon both of which the quantity of energy depends, must be taken into consideration. One of these is called the *capacity factor*, the other the *intensity factor*. The energy is found to depend upon the product of these.

For example, in the expression for mechanical kinetic energy, $K = \frac{1}{2}MV^2$, M is the capacity factor, V^2 the intensity factor. In the case of the energy of a compressed gas the volume is the capacity factor, the pressure the intensity factor. In the case of heat energy, the heat-capacity at a certain temperature is the capacity factor, the temperature the intensity factor. In the case of an electric charge, the electrical quantity is the capacity factor, its potential the intensity factor.

It is the intensity factor which determines the tendency towards change of energy either as to distribution or form. It is because of the pressure of compressed gas that it tends to expand. The temperature of a mass determines the tendency towards a transfer of heat; the potential of an electric charge determines the tendency toward a flow of electricity.

The intensity factor is a relative quantity. Thus gaseous pressure, temperature, electric potential are magnitudes relative to some value taken as a standard. Hence neither transfer nor transformation of energy can take place unless there exists a difference in the intensity factor between different parts of the system. Thus there is no transfer of gas between two connecting reservoirs if the pressure of the gas is the same in each; there is no resultant transfer of heat between two bodies whose temperature is the same; nor of electricity between two bodies at the same potential.

63. Function of Machines.—It will be seen from what precedes that the function of a machine is merely to transfer or transform energy, not to create it.

Neglecting friction and other harmful resistances, the effective force applied to any machine multiplied into the distance through which its point of application is moved in the direction in which the force acts, is equal to the resistance overcome multiplied into the distance through which it is overcome. That is, the mechanical work supplied to the machine is equal to the mechanical work done by the machine in overcoming resistance.

On account of friction and various other harmful resistances, the latter of these is in practice always less than the former.

EXAMPLES: Lever, wheel and axle, pulley, and other "mechanical powers."

The *efficiency* of a machine is the ratio of the work which it does to the work done upon it.

64. Measurement of Work of Machinery.—The mechanical energy expended in driving a machine may be measured by different forms of dynamometer or erg-meter.

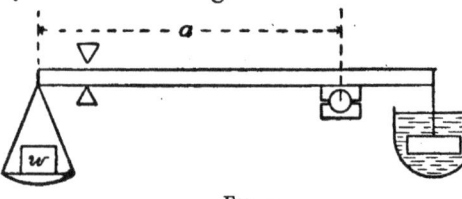

FIG. 7.

1. Friction Brake. The shaft (Fig. 7) is made to run at its normal speed, overcoming the friction of the brake as a resistance. Calling this speed n ($r.p.m.$), w the balancing weight in the brake-pan, a the arm of the brake, the work done by the machine is $W = w\,a \times 2\pi n$ per minute, whence $H.\,P. = \dfrac{2\pi n w a}{33000}$, if w and a are expressed in pounds and feet respectively.

2. Transmission Dynamometer. The work done is measured under actual conditions of use. In the torsion dynamometer the torque is determined from the twist of a shaft or helical spring through which the power is transmitted.

65. Perpetual Motion.—In its technical meaning, a "perpetual-motion machine" is not a machine which would run indefinitely if its working parts remained unimpaired, but a machine which does work without the expenditure of energy.

Thus a windmill operated by the trade-winds, or a turbine operated by Niagara Falls, or an engine working automatically by the rise and fall of mercury in a thermometer or barometer tube would not be a perpetual motion machine, since energy would be expended in its operation. Nor is the revolution of a planet about the sun perpetual motion since no work is done in the revolution.

It is a logical consequence of the Principle of the Conservation of Energy that "perpetual motion" as above defined is an impossibility.

66. Sources of Energy.— TAIT.

   a. Potential.
       1. Fuel.
       2. Food of Animals.
       3. Ordinary Water Power. (Head.)
       4. Tidal Water Power.
   b. Kinetic.
       5. Winds and Ocean Currents.
       6. Hot Springs and Volcanoes.

IMMEDIATE SOURCES.
       1. Primordial Energy of Chemical Affinity.
       2. Solar Radiation.
       3. Energy of Earth's Rotation.
       4. Internal Heat of Earth.

67. The Sun as a Source of Energy.

—Substantially all the energy which is utilized by man is derived from the sun, *e. g.*, that of water power, wind power, fuel, food, electrical power, though not that of tidal water power.

The heat entering the earth's atmosphere from the sun according to the determination of Abbot (1908) is 2.1 calories (gram-degrees) per square centimeter per minute. This is known as the *solar constant.* Hence energy is received at the rate of 0.15×10^7 ergs per second = 0.15 watts. This corresponds to radiation from each square centimeter of the sun's surface at a rate of about 9 H. P. The temperature of the sun's surface is probably in the neighborhood of 6,000° to 7,000° C.

Helmholtz (1854) explained the origin of this energy as due to slow contraction of the gaseous sun. A yearly diametral contraction of not over 250 feet would suffice to produce the observed amount. Assuming the sun to have contracted from a sphere filling the orbit of Neptune the calculated possible radiation would not suffice for over 18,000,000 years of emission or thereabouts. On these data Kelvin has based an estimate of the maximum possible "age of the earth." Recent discoveries within the domain of radioactivity have somewhat disturbed these conclusions.

68. Historical.

—The truth of the doctrine of the Conservation of Energy in its complete form came to be recognized very gradually, principally during the first half of the nineteenth century. No one person can be named as its originator. Newton was apparently cognizant of the principle so far as concerns machines, and recognized the fact that there could be no gain of power by such means. The great French masters of applied mathematics in the latter part of the eighteenth century, D'Alembert, Lagrange, Laplace, and others, assumed its truth so far as concerns mechanical forces, and based their most important generalizations upon it.

Early in the nineteenth century Dr. Thomas Young introduced the term "energy" to denote mechanical energy or more strictly *vis-viva*.

In 1798 Benjamin Thompson, Count Rumford, from a study of the enormous amount of heat developed in boring a cannon showed conclusively that heat could not be an "imponderable fluid," as was then believed, but, on the contrary, is a "mode of motion," and produced by friction. It was not until a much later date, however, that this view came to be the accepted one.

Rumford's views were confirmed and adopted by Sir Humphry Davy (1799) and Dr. Thomas Young (1807), but beyond these they met with slight acceptance. Even Fourier (1822) and Carnot (1824) accepted the "caloric" theory of the material nature of heat.

Séguin in France (1839), Mayer in Germany (1842), Colding in Denmark (1843), and Helmholtz in Germany at about the same time, all adopted the view that heat is a form of what we now call energy, and endeavored to ascertain the laws of its generation and transformations. Mayer showed a way of computing the mechanical equivalent of heat from already known data, which though subject to criticism from a logical point of view nevertheless proved capable of furnishing approximate results.

Joule in England, whose work began as early as 1840, made a series of contributions to this branch of science which are of inestimable value. In 1843 he showed that the heat produced in a given circuit by a given quantity of electricity generated by a magneto-electric machine was the same in amount as that which would be produced by direct transformation into heat of the work consumed in producing the current of electricity by the machine. He also obtained from his measurements a value for the mechanical equivalent. Shortly thereafter he determined this constant by direct measurement under different conditions of the work done by friction and of the resulting heat. Using various methods of producing heat by the expenditure of mechanical work, all of which were found to give a like value of the mechanical equivalent, he showed this to be independent of the particular manner in which the work was done. In 1850 he published a paper of great completeness giving a value of $J = 772$ foot-pounds which continued to be accepted for almost thirty years.

In 1847 Joule in Manchester and Helmholtz in Berlin each read a paper now regarded as epoch-making in importance, in which the principle of the conservation of energy in its full breadth was clearly set forth. But their views were not accepted in either country, and Poggendorff's *Annalen*, the leading physical journal in Germany, declined to publish the paper of Helmholtz. But largely through the ability and influence of William Thomson (afterwards Lord Kelvin) and Du Bois Reymond, the new views rapidly gained acceptance.

The reverse transformation of heat into mechanical work was investigated by Hirn, Rankine, Kelvin, Clausius, and others, showing the same quantitative relationship to exist between heat transformed and the resulting work.

The expression "Conservation of Energy," replacing the earlier and confusing expression "Conservation of Force," was introduced by Rankine.

The laws of relationship of electrical quantity and chemical decomposition were discovered by Faraday (1833), those relating to the heating effects of electricity by Joule (1841). More recently the laws relating to heat produced by chemical action have been determined by Favre and Berthelot.

All the results obtained, without exception, have been such as to confirm and extend the conclusion that energy is neither created nor destroyed in any of its varied transformations.

69. Dissipation of Energy.—*All energy tends to pass from a higher form to a lower one, and ultimately to assume the form of uniformly diffused heat.*—KELVIN.

Hence, although the total energy of a system remains constant, the amount of its *available* or *free* energy continually diminishes toward zero. Thus it is easy to transform completely the total amount of mechanical energy of a system into heat energy, but it is impossible to transform the total amount of heat energy of a system into mechanical energy. The total electrical energy of a system can be transformed into heat, but the total heat energy of a system cannot be transformed into electrical energy.

For example, it can be shown that with a perfect steam or other heat engine, if the vapor enters the cylinder at a temperature of 200° C., and leaves it at a temperature of 110° C., the efficiency can be only 19 per cent. With the actual engine it is much lower.

70. Application of Principle to Cosmical Phenomena.— The principle of the dissipation of energy applies to the physical universe as far as we know it, whence we conclude that the available energy of the universe tends toward zero. There is, however, no evidence of retardation of planets or comets such as would be caused by a resisting medium in space.

71. Transmission of Energy.—1. Transference of body possessing energy. 2. Waves.

72. Waves.—In any wave the wave-form alone progresses, while the particles at any point merely assume an oscillatory or rotary motion.

Waves are of three kinds: *longitudinal, transverse,* or *torsional,* according as the motion of the particles is parallel to the line of transmission of the wave, transverse to it, or twisting about it as an axis.

The following classes of wave may be produced in ordinary matter: (a) Gravitational waves in liquids, due to displacement of level. (b) Capillary waves or surface ripples, due to surface tension of the liquid. (c) Pressural waves in body of solid, liquid or gas, due to volume elasticity. (d) Distortional waves in solid, due to elasticity of shearing, bending, twisting. Of these (a) and (b) are necessarily transverse waves, (c) longitudinal, (d) either transverse or torsional.

Water waves are transverse; the sound wave is longitudinal.

In the ether, electromagnetic waves exist, of which light waves and the electric waves, utilized in space telegraphy, are examples. These are due to electrical and magnetic properties of the ether which are analogous to elasticity and mass in ordinary matter. Electromagnetic waves are transverse in their character.

CURVILINEAR MOTION. ROTATION.

73. General Principles. Definitions.—Curvilinear motion is produced by the continuous action of a deflecting force upon a moving particle or mass. The deflecting force may vary in any manner whatever, both as to magnitude and direction.

An important case is that in which the deflection is always toward the same point, in which case the deflecting force is called a *centripetal force*. Such a point is called a *center of force*. The so-called "tangential or projectile force" in reality is not a force, but the momentum of the particle.

It will readily be seen that the nature of the orbit traversed by the particle will depend upon the manner in which the magnitude and direction of the deflecting force vary. Hence from the form of the orbit the law of variation of the deflecting force can be determined.

Newton showed that if the orbit of a body is a conic section the line of action of the deflecting force must always pass through a focus of the curve, and furthermore must vary in magnitude from point to point of the curve in the inverse ratio of the square of the distance of such point from that focus; and conversely. This is the case with all members of the solar system. All the planets move in ellipses of small eccentricity with the sun situated in one focus of the ellipse. Many comets also move in orbits which are elliptical, but of great eccentricity. Most comets move in sensibly parabolic orbits, and some, apparently, in hyperbolic orbits.

Since the centripetal force acts to draw the revolving body inward toward the center of force, it follows from the 3rd Law of Motion that this center sustains a radially-outward pull equal in magnitude to the centripetal force. It is this tendency of the center of motion to move radially outward which led to the conception of the definite existence of a *centrifugal force*. In fact, there is no such force. It is only the reaction on the axis of the centripetal force, which pulls the body and the center equally towards one another, though the fixation of the center prevents this from moving. The revolving body always tends to move tangentially in virtue of its inertia.

As the tangential momentum, however, tends to increase the radius of revolution, the term "centrifugal force" has come to be habitually used by engineers and others as denoting a component of the momentum which is opposite to the centripetal force, and, necessarily, equal to it.

74. Centripetal Force in Circular Orbit.—The case most frequently occurring in the ordinary applications of Physics is that of a particle or body revolving in a circle about a center of force situated at the center of the circle. The curvature of the circle being constant, the centripetal force in such an orbit is of constant intensity. Denote it by F.

To determine its value let C (Fig. 8) be the center of the circle in whose circumference the particle revolves, PR the space over which the particle would pass in time t in virtue of its tangential velocity, PM the space which the particle would traverse in the same time in virtue of the action of the centripetal force alone. Let PQ be the chord of the arc in which the particle moves. From the geometry of the circle, $PQ^2 = PM \times PD$.

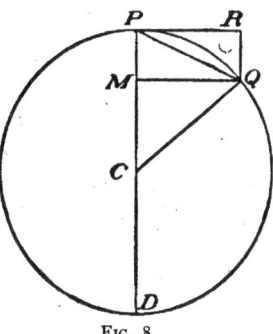

FIG. 8.

Denote by f the centripetal acceleration, and by v the constant orbital velocity of the particle. If s is the space which would be traversed toward C in time t under the influence of the centripetal force, $\dfrac{d^2s}{dt^2} = f$. But $s = PM = \dfrac{PQ^2}{PD}$. Also for indefinitely small arcs, the chord and arc coincide, whence $PM = s = \dfrac{v^2t^2}{2R}$ and

$$\frac{d^2s}{dt^2} = \frac{v^2}{R} = f. \quad \text{Hence } F = Mf = M\frac{v^2}{R}.$$

Hence, *the centripetal force varies directly as the mass of the particle and the square of the velocity, and inversely as the radius of the orbit.*

Evidently M may be written in either gravitation or absolute units, in which case F will be expressed in similar units.

The centripetal force of any extended body or system of bodies is the same as if the total mass were concentrated at its center of mass. Hence the preceding demonstration can be applied immediately to such cases.

The formula may also be put into the following form:—

Let T be the time of revolution; then $vT = 2\pi R$, and $v = \dfrac{2\pi R}{T}$.

Hence $F = M\dfrac{v^2}{R} = M\dfrac{4\pi^2 R}{T^2}$. Hence, for bodies revolving in circular orbits, *the centripetal force varies directly as the mass and the radius of the orbit, and inversely as the square of the time of revolution.*

75. Velocity in Orbit.—The uniform velocity with which a body revolves in a circular orbit is equal to that which the centripetal force would generate by its constant action upon the body through half the radius of the orbit. For

$$f = \frac{v^2}{R}; \text{ whence } v = \sqrt{fR}.$$

Suppose the centripetal force to impel the body from a state of rest until it attains the velocity v. Call s the space described. Then

$$v = \sqrt{2fs}; \text{ whence } \sqrt{fR} = \sqrt{2fs}, \text{ and } s = \frac{R}{2}.$$

76. Body Revolving in Vertical Orbit.—That a body may revolve in a vertical circular orbit, its centrifugal force must at least be equal to its weight. To find the minimum velocity requisite for this, put $W = F$ in the general equation. Then

$$W = \frac{Wv^2}{gR}, \text{ and } v = \sqrt{gR}.$$

ILLUSTRATION: Centrifugal Railway.

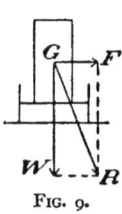

FIG. 9.

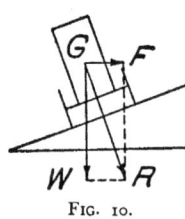

FIG. 10.

77. Practical Illustrations.—Centrifugal dryer, blower, pump, cream-separator; centrifuge; bursting of fly-wheels; vehicle on curved road (Figs. 9 and 10); depression of inner rail on curve.

78. Conical Pendulum.—In the conical pendulum the ball P must assume such a position that the cord AP is in the line of action of R, the resultant of the weight and the centrifugal force of the ball.

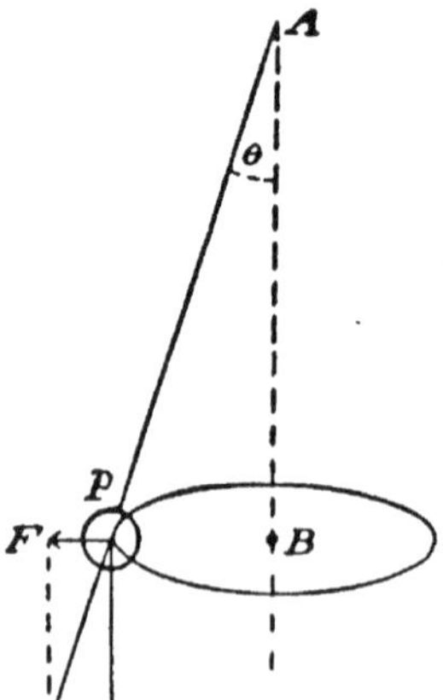

Under these circumstances $\tan \theta = \dfrac{F}{W} = \dfrac{r}{h}$,

denoting by r the radius of the circle of revolution of P and by h the altitude of the cone described by the suspending cord. The time of rotation is

$2\pi\sqrt{\dfrac{h}{g}}$. For the centrifugal force $F = W\dfrac{r}{h} =$

$\dfrac{W}{g}\dfrac{4\pi^2 r}{T^2}$, and $T = 2\pi\sqrt{\dfrac{h}{g}}$.

FIG. 11.

APPLICATIONS.—Watt's governor for steam engine; regulating mechanism for chronographs; driving clock for equatorial telescope.

79. Equilibrium Surface of Liquid rotating about Vertical Axis.—This is a paraboloid of revolution. At every point on the free liquid surface that surface must be at right angles to the resultant of the weight W of the particle and its centrifugal force F.

Hence (Fig. 12) $\tan \theta = \dfrac{W}{F} = W \div \dfrac{W 4\pi^2 r}{g T^2} =$

$\dfrac{g T^2}{4\pi^2 r}$; and $r \tan \theta = \dfrac{g T^2}{4\pi^2}$ which value is constant for all points on the liquid surface. Hence $r \tan \theta = AC$, the subnormal of the curve, is a constant, which is a well-known property of the parabola.

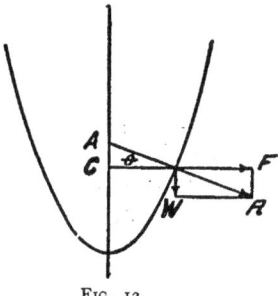

FIG. 12.

80. Centrifugal Force at Equator.—The magnitude of the centrifugal force at the equator due to the earth's rotation is found by substituting the proper values of R and T in the general equation.

The equatorial radius of the earth $= 6378250$ meters (Clarke, 1880), and the value of g at the equator as measured is 9.78 meters. Hence

$$F = M \frac{v^2}{R} = W \frac{4\pi^2 R}{g T^2} = W \frac{4 \times (3.141593)^2 \times 6378250}{9.78 \times (86164)^2} = W \frac{1}{288.4}.$$

Hence the effect of the centrifugal force at the equator is to diminish the weight of every body by its $\frac{1}{288}$ part. Since F varies as v^2, a velocity of approximately 17 times the present would generate a centrifugal force at the equator entirely counterbalancing the action of gravity.

81. Variation of Centrifugal Force with Latitude.—It will appear from a consideration of Fig. 13 that the value of the centrifugal

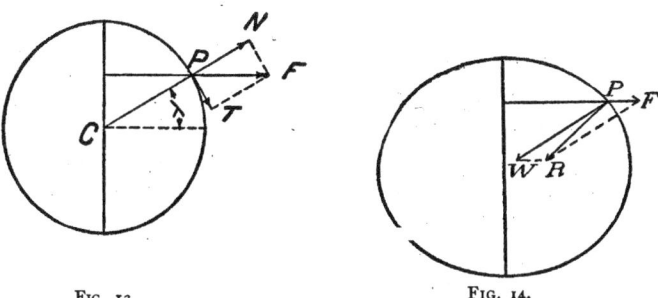

FIG. 13. FIG. 14.

force in latitude λ, represented by PF, will be to its value at the equator in the ratio of the respective radii of revolution of a point in these latitudes. Hence $F_\lambda = F_0 \cos \lambda$.

The diminution of weight due to centrifugal force in latitude λ is represented by the normal component PN in the diagram. Calling this N, we have $N = F_\lambda \cos \lambda = F_o \cos^2 \lambda$.

It will also be seen that the existence of a tangential component, PT, of the centrifugal force acts to produce a deviation of the plumb-line toward the equator. For latitude $45°$ the value of this can be shown to be $11'\ 30''$.

Other important phenomena in which the centrifugal force due to the earth's rotation is concerned are those of cyclones and storms in general, and the tides. A suggested deviation of river courses is doubtful.

82. Effect of Centrifugal Force on ·Figure of the Earth. Spheroid of Equilibrium.—The tangential component PT acts to produce motion in a movable body, as, for example, the water of the ocean, from the poles toward the equator. Hence with a fluid mass, such as the earth must have been at a former remote period, there would result a heaping up of matter in the equatorial regions, so that the earth thus rotating could no longer be spherical.

The actual form assumed by such a fluid rotating mass must be such that the surface at every point is normal to the resultant of the weight of the particle and the centrifugal force. If P, Fig. 14 (p. 35), is any such point, PR, the resultant of the weight PW and the centrifugal force PF must be normal to the surface at P.

Maclaurin (1740) showed that an oblate spheroid fulfils this condition. It has since been shown by Jacobi and Poincaré that there are several other possible figures of equilibrium.

From the mass of the earth and its period of rotation it is possible to calculate the amount of the polar flattening.

The ellipticity of the earth as measured is $\frac{1}{300}$. That of Jupiter, whose period of revolution is only 9 h., 55 m., is $\frac{1}{17}$.

83. Plateau's Experiment.—Sphere of oil immersed in mixture of alcohol and water of same density, becomes spheroidal when rotated, and finally throws off equatorial rings if the velocity is increased.

84. Theories of Evolution of Solar System.—(a) Nebular Hypothesis of Kant (1755) and Laplace (1796). Difficulties in the way of its acceptance. (b) Planetesimal or Spiral-Nebula Hypothesis of Chamberlain (1905).

85. Kelvin's Estimate of "Age of Earth."—Ellipticity (assumed to be substantially unchanged since solidification) in connection with increase in length of day of 22 seconds per century from tidal retardation leads to the conclusion that some 20,000,000 years have elapsed since solid crust was formed. Estimates based on an entirely different class of data lead to a value of the same order of magnitude, as has already been stated in connection with the energy of solar radiation.

OBJECTIONS.— Plasticity of earth has probably allowed a subsequent change in ellipticity. Amount of retardation is not certain.

86. Rotation of Rigid Bodies.—The effect of an unbalanced centrifugal force is two-fold, tending (1) to shift the axis of rotation as a whole, *i. e.*, to produce a translatory motion of the axis, and (2) to produce angular deviation of the body. (See Figs. 15, 16.)

87. Free Axes. Principal Axes.— It will be seen from the figures that, if the axis of revolution passes through the center of mass, the tendency to translatory motion disappears. That there may also be no tendency to angular motion, the axis of rotation must coincide with an axis of symmetry of the body. (See Fig. 17.) If

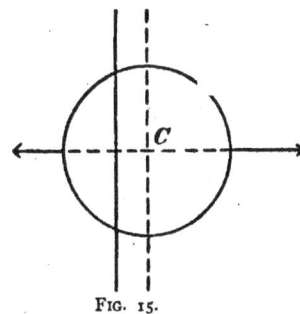

FIG. 15.

the axis of rotation is parallel to an axis of symmetry, but does not pass through the center of mass, there will evidently be no tendency to angular deviation. An axis about which a body may revolve without causing any tendency to angular deviation is called a *principal axis.* Any axis

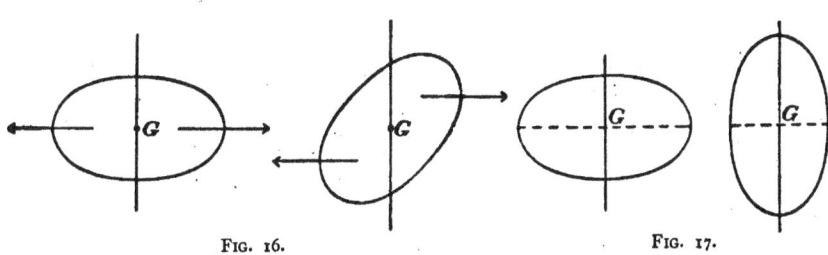

FIG. 16.   FIG. 17.

about which a body may revolve without producing a tendency to either angular deviation or translation of the axis of rotation is called a *free axis.* The free axes are evidently principal axes passing through the center of mass.

In an ellipsoid with three unequal axes, these are the free axes. In a right elliptical cylinder the free axes are the axis of the cylinder and the major and minor axes of the elliptical middle section of the cylinder. Any diameter of a sphere is a free axis; also, any diameter of the equator of an oblate or prolate spheroid, together with its polar axis.

PRACTICAL APPLICATIONS IN MACHINERY: Balancing of fly-wheels and other rotating pieces; balancing of crank on driving-wheel of loco-motive.

88. Axis of Stable Rotation.—An inspection of Fig. 16 will also show that a body is in stable equilibrium only when rotating about its shortest diameter.

In machinery several practical considerations frequently require that a rotating piece shall be rotated about a longer rather than a shorter axis of symmetry. This, however, is not a case of free rotation, and the rigidity of the shafting is made such as to counterbalance any centrifugal couple that may be generated.

89. Analogies between Translatory and Rotary Motion.—Some further characteristics of the rotary motion and energy of rigid rotating bodies are most readily considered in this connection.

It is easily seen that the effect of a constant unbalanced torque is to produce a uniformly accelerated rotation in a mass, and hence that relations hold between α, ω and θ identical with those which obtain in the case of uniformly accelerated translatory motion.

Denote the angular velocity of such a body by ω, its angular acceleration by α, and the angle described in time t under the action of the constant torque by θ. Then

$$\frac{d\theta}{dt} = \omega, \qquad \frac{d\omega}{dt} = \frac{d^2\theta}{dt^2} = \alpha.$$

From this it follows that

$$\omega = \alpha t, \qquad \theta = \tfrac{1}{2}\alpha t^2, \qquad \omega = \sqrt{2\alpha\theta},$$

which formulæ are of the same character as those already proved for uniformly accelerated translatory motion; viz.:—

$$v = at, \qquad s = \tfrac{1}{2}at^2, \qquad v = \sqrt{2as}.$$

It will furthermore be seen that a simple relation exists between the work done upon a rotating mass by a torque T and the resulting kinetic energy; i. e., $T\theta = \tfrac{1}{2}\omega^2 I$. For calling F, r, respectively, the force and arm of the torque, we have $T\theta = Fr\theta = FS = \tfrac{1}{2}\omega^2 I$. (See § 46, p. 21.)

It will be noted that the moment of inertia, I, takes the same place in the dynamics of rotation that the mass M takes in the dynamics of translation. This will be apparent from the comparisons in the following table:—

TRANSLATION	ROTATION
$v = at$	$\omega = \alpha t$
$s = \tfrac{1}{2}at^2$	$\theta = \tfrac{1}{2}\alpha t^2$
$v = \sqrt{2\,as}$	$\omega = \sqrt{2\alpha\theta}$
$F = Ma$	$T = I\alpha$
$Ft = MV$	$Tt = I\omega$
$K = \tfrac{1}{2}MV^2$	$K = \tfrac{1}{2}I\omega^2$
$FS = \tfrac{1}{2}MV^2$	$T\theta = \tfrac{1}{2}I\omega^2$

90. Ballistic Pendulum.—A heavy block of wood is suspended on knife-edges. A bullet whose velocity is to be measured, is fired into it, in a line at right angles to the axis of suspension, producing an angular

·velocity ω. The deflection of the pendulum is measured and ω determined from this. Calling m the mass of the bullet, v its velocity, I the moment of inertia of the pendulum, k the distance from the axis of suspension to the line of entrance of the bullet, $I\omega = m\,v\,k$ and $v = \dfrac{I\omega}{mk}$. In strictness the mass of the bullet should be taken into account in the value of I, but this is so small that practically it is unimportant.

In order not to jar the support of the pendulum the line of fire should pass through its *center of percussion* as will be explained later.

The moment of inertia may either be calculated or determined experimentally.

91. Moment of Momentum. Angular Momentum.—The moment of a force F relatively to any point O is Fr_0 where r_0 is the normal drawn from O to the line of action of F. In like manner if a mass moves with a velocity v, its moment of momentum relatively to O is $T_0 = mvr_0$, r_0 being normal to the direction of v. This quantity, also called angular momentum, represents the rotary effect produced by the momentum mv.

92. Conservation of Angular Momentum. — The angular momentum of a system of bodies is not in any way altered by the mutual actions of the masses composing that system.

Let m_1 m_2, Fig. 18, be two masses, and let their angular momenta be taken relatively to an axis passing through any point, as O, normal to the plane of the paper. The action of m_1, m_2 on each other will be along the line joining them. As action and reaction are equal and opposite the force F_1 with which m_1 is drawn toward m_2 will equal the force F_2 by which m_2 is drawn toward m_1.

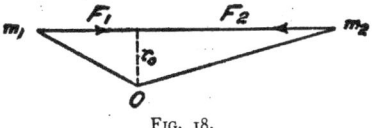

FIG. 18.

Under the action of this mutual attraction there will be generated in each in time t a quantity of momentum such that $Ft = m_1v_1 = m_2v_2$. But as will be seen from the figure $m_1v_1r_0$ is the angular momentum of m_1 relatively to O and $m_2v_2r_0$ that of m_2 relatively to the same point. Since these are in opposite directions the resultant angular momentum $= 0$. Hence they can in no way alter the previously existing angular momentum of the system.

What is true of two masses m_1, m_2 is equally true of any number, for the effect of any one mass upon any other in the system can be dealt with in the same manner, and since the effect of every individual mass upon every other is to produce equal and opposite angular momenta with therefore a resultant of zero, the total mutual effect of the masses composing the system must also be zero.

93. Conservation of Areas.—It follows from the preceding demonstration that if we suppose a rotating mass m_1 with radius r_1 and velocity v_1 so to move that its radius changes to r_2 the velocity will become v_2, having a value such that $m_1v_1r_1 = m_1v_2r_2$. Hence considering a very brief time dt, we have $r_1v_1dt = r_2v_2dt$. But the first member of this equation is double the area of the elementary triangle described in time dt by the mass when its radius is r_1 and the second double the area described by it when its radius has become r_2. Hence the areas traversed by the radii vectores in equal times are the same.

Because of this fact the conservation of angular momentum has sometimes been called the " conservation of areas." Kepler's Second Law is an illustration of it.

It will be seen from what has been said that if a portion of a revolving mass is transferred from a point near the axis of revolution to one farther removed from it the angular momentum of the mass is diminished by the same amount as that by which the angular momentum of the portion moved is increased. And while the movable part is gaining speed it reacts on the mass as a whole.

94. Composition of Rotations.—If the angular velocities of two rotations are represented by the two adjacent sides of a parallelogram the diagonal of that parallelogram will represent the angular velocity of the resultant.

Let AB, AC (Fig. 19) represent component angular velocities ω_1, ω_2. Consider any point P whose coördinates referred to AC, AB are x,y.

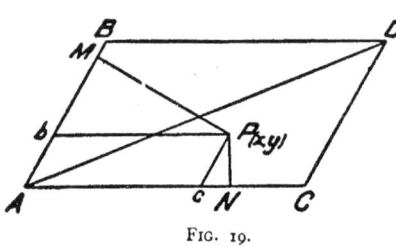

FIG. 19.

Draw PM, PN normal to AB, AC. The downward linear velocity of P due to the rotation AB is $\omega_1 PM = \omega_1\, x \sin BAC$. The upward linear velocity of P due to the rotation AC is $\omega_2 PN = \omega_2 y \sin BAC$. For points on the resultant axis these opposite rotations must be equal. Placing $\omega_1 x \sin BAC = \omega_2 y \sin BAC$, we have $\dfrac{x}{y} = \dfrac{\omega_2}{\omega_1} = \dfrac{AC}{AB}$, which is the equation of the diagonal AD. This diagonal therefore represents the resultant axis in direction.

AD furthermore represents the resultant rotation in magnitude. For, consider the motion of the point C. Its linear velocity due to the resultant rotation must be equal to that due to the simultaneous action of the two components. But the velocity due to rotation $AC = 0$. That due to $AB = AC \sin BAC \times \omega_1$. That due to $AD = AC \sin DAC \times \omega_r$, calling ω_r the resultant angular velocity. Hence as these last must be equal, $AC \sin BAC \times \omega_1 = AC \sin DAC \times \omega_r$ or

$$\omega_r = \omega_1 \frac{\sin BAC}{\sin DAC} = AB \frac{\sin BAC}{\sin DAC} = AD.$$

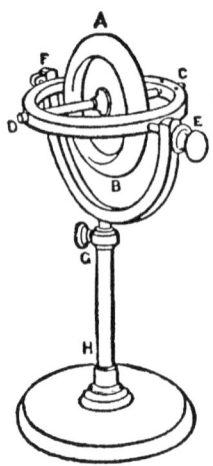

FIG. 20.

95. Gyroscope or Gyrostat.—Referring to Fig. 20, let DC, the axis of spin of the disk, be designated by X, the axis FE by Y, the vertical axis HG by Z. Assume that there is perfect freedom of motion about each of these axes. With the instrument balanced so that its center of mass lies at the center of the rotating disc AB, the axis of spin remains parallel to itself in whatever manner the instrument as a whole is moved. If, however, the instrument is unbalanced, as e. g., by hanging a weight from D, a precessional rotation takes place about Z, instead of rotation about Y such as would occur under like circumstances were the disc at rest.

The precessional motion results from the combination of two rotations, the spin of the disc about X and the rotation about Y due to the couple produced by the suspended weight.

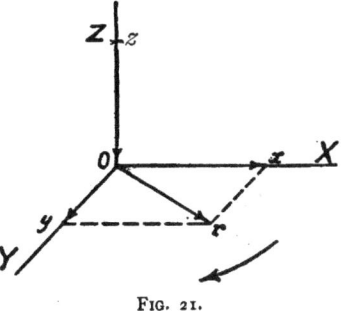

FIG. 21.

Assuming the direction of spin to be right handed as seen from D we may represent the angular velocity about DC by Ox, Fig. 21.

The angular velocity due to the couple produced by the suspended weight may be represented by Oy. Then $zO = Or$, will represent the resultant couple, in magnitude and direction. The axis of spin will therefore move in such manner as to tend to approach Or, as indicated by the arrow in $X O Y$, remaining, however, in a horizontal plane, rotating about $O Z$, Fig. 21, or $H G$, Fig. 20. But as this rotation carries the axis $F E$ with it, the direction of Oy and hence of Or is constantly changing, performing a rotation about $O Z$; whence the precessional movement is continuous.

It is easily seen from the construction of Fig. 21 that if either the direction of spin about DC or that of the gravitational couple about FE is reversed the direction of precession will likewise be reversed.

The following is a general explanation on dynamical principles of the phenomena just considered. Fig. 22 represents a disc rotating right-handedly about an axis through O perpendicular to the plane of the paper as an axis of spin. Suppose that a second rotation is given to the disc, about YOY, the upper half moving toward the eye. The particles in quadrant 2 on account of the spin about O are moving upwards and increasing their distance from YOY. Hence on account of their momentum their effect is to produce a tendency in 2 to move away from the observer.

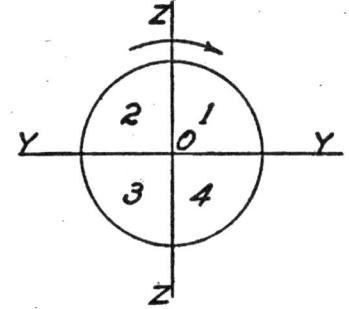

FIG. 22.

In quadrant 1 on the other hand the particles are descending towards YOY, and on account of their momentum the diminution in their radii of revolution about YOY will cause a pressure throughout 1 towards the eye. In likè manner in quadrant 4 there will be a pressure toward the eye and in quadrant 3 away from it. Hence a rotation will be produced about ZOZ as an axis.

The action described above is analogous to the case of a ball whirled by a string, which when the string is lengthened so that its radius of revolution is increased has its velocity correspondingly slackened, and which if it were being pushed forward would press backward in opposition. The conception of the action may also be helped by supposing the disc to be a rotating disc of liquid contained in a flat box, mounted like the gyroscope.

The several actions take place in accordance with the principle of the conservation of angular momentur. (See § 92, p. 39.)

The great resistance opposed by a gyroscope to sudden angular deviation of its axis of rotation is due to the action just explained. The combination of couples produces a resultant couple such as would generate a precessional motion, and if the impressed force is such as to prevent this, the resisting effect due to the momentum of the wheel becomes very great.

Other gyroscopic phenomena are to be explained according to the principles just laid down.

Thus with diminishing speed the precession is faster because Ox becomes less in proportion to Oy. If the weight on D is increased or if this point is pushed down the rate of precession is increased; if pushed up, diminished. If the axis of spin is tilted the precession is faster as the effective value of Ox is less. If the precession is accelerated by external force, the center of mass of the system rises; if the precession is retarded, the center of mass descends. The explanation of this may be seen by combining a rotation about the axis OZ with Or. If the axis HG is fixed, so as absolutely to prevent precession, the added weight will produce rotation about FE, as if the disc were not rotating. There will be a torsional stress, however, in the vertical axis due to the precessional tendency. Retardation of precessional motion by friction of the axis HG will cause a gradual depression of the axis of spin.

96. Phenomena and Applications of Gyroscope.—Rolling hoop; bicycle; game of diabolo; rifled guns; steadying gyroscope on vessel to prevent rolling; steadying gyroscope and steering apparatus in Howell torpedo; Obry's steering apparatus in Whitehead torpedo; Brennan's mono-rail system. Also various devices proposed for determining latitude at sea; for transferring directional line, as, e. g., the meridian, from surface to bottom of mme, etc., but none of them as yet practical.

97. Astronomical Gyroscopic Phenomena.—Parallelism of axis of earth to itself, causing phenomena of seasons. Precession of the equinoxes,

caused by action of sun on equatorial protuberance of earth, with period of about 26,000 years. Nutation, caused by action of moon on equatorial protuberance, with period of about nineteen years. Periodic variations of latitude investigated by Chandler, cause uncertain.

Application of principle of gyrostat by Kelvin to explain hypothesis of elastic-solid luminiferous ether. Kelvin's hypothetical gyrostatic elastic atom.

98. Top.—The various precessional phenomena of the top are to be explained upon the same principles. The rise of a top from an inclined to a vertical position is due to the combination of precessional motion with the rolling motion of the peg. The roll tends to carry the top along faster than the precession and so to accelerate this, thereby causing the center of mass of the top to rise.

Maxwell's dynamical Top. Newton's geometric Top. Griffin's stone-pulverizer.

PERIODIC MOTION.

99. Definitions.—A particle possesses a periodic motion when it successively traverses the same path, returning to any given point after having passed through a complete cycle of changes. Such motion may take place in a closed curve, an arc, or a straight line. The time occupied in completing one cycle is called the *period*.

The earth revolves about the sun in an elliptical orbit, with a period of one year; a "seconds" clock-pendulum vibrates in a circular arc, with a period of two seconds; a body suspended by an elastic cord executes a vibrating periodic motion in a vertical line, with a period equal to the time occupied by one complete "up and down" motion.

Any periodic motion can be represented by a curve constructed with times as abscissas and the corresponding displacements of the particle as ordinates. Such a curve is evidently periodic. When thus constructed it is a *curve of spaces* or *displacements*.

A curve representing periodic motion may also be constructed in which times are abscissas and the corresponding velocities are ordinates. This is a *curve of velocities*.

The curve of velocities can always be derived from the curve of spaces, and conversely, since $v = \dfrac{ds}{dt}$.

100. Simple Harmonic Motion.—An extremely important case of periodic vibratory motion is that known as *simple harmonic motion*.

Referring to Fig. 23, let us suppose a particle P to revolve uniformly in the circle there shown, whose radius is r, and likewise that a second particle

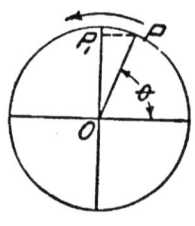

FIG. 23.

P_1 moves to and fro along the vertical diameter of that circle in such manner as always to coincide in position with the projection of P on that diameter. The motion of the vibrating particle is a simple harmonic one.

The position in its path of the vibrating particle at any instant is denoted by what is called the *phase* of the vibration. This term is defined more precisely as follows: "The phase of a simple harmonic motion at any instant is the fraction of the whole period which has elapsed since the moving point last passed through its middle position in the positive direction." The phase is measured by the angle θ in Fig. 23.

When a particle after displacement executes a vibration of this character the restoring force at each point must be proportional to the displacement.

The acceleration at any point P_1 is given by the equation $\dfrac{d^2y}{dt^2} = a$. But $y = r \sin \theta = r \sin \omega t$, calling ω the angular velocity and t the time corresponding to the phase-angle θ; that is, the time occupied by the vibrating particle in passing from O to P_1. Then $\dfrac{dy}{dt} = \omega r \cos \omega t$ and $\dfrac{d^2y}{dt^2} =$ $- \omega^2 r \sin \omega t = - \omega^2 y$; whence $a \propto y$, the displacement of the particle; and the same must be true of the restoring force, since $F = Ma$. The opposite sign of a and y indicates that the acceleration is opposite in direction to the displacement.

The converse of the proposition is likewise true, since this law of force can produce only one mode of variation of the acceleration.

If T be the time of one complete revolution of P and hence also the time of one complete vibration of P_1, $T = \dfrac{2\pi}{\omega}$ and $\omega = \dfrac{2\pi}{T}$.

It has just been shown that $a = \omega^2 y$ (dropping the minus sign as indicative only of direction) whence $\omega^2 = \dfrac{a}{y} = \left(\dfrac{2\pi}{T}\right)^2$ and

$$T = 2\pi \sqrt{\dfrac{y}{a}} = 2\pi \sqrt{\dfrac{displacement}{acceleration}}.$$

It also appears from this expression that *simple harmonic vibrations are isochronous*, since the ratio of y to a in any particular case is a constant and independent of r.

101. Equations of Harmonic Curves.—It is apparent that the movement of a particle possessing a simple harmonic motion is represented by the equation $y = r \sin \theta$, or $y = r \sin \omega t$, where $\omega = \dfrac{2\pi}{T}$, y being the displacement. Hence, denoting times by abscissas, the equation of the curve of displacement is $y = r \sin \omega x$, which is the equation of a sinusoid, r being the amplitude of vibration.

The corresponding curve of velocities is readily found. Since $v = \dfrac{ds}{dt}$ and in this case $s = y = r \sin \omega t$, $v = \omega r \cos \omega t$. Hence, taking velocities as ordinates and times as abscissas, the equation of this curve is $y = \omega r \cos \omega x$. This is also a sinusoid, but displaced along the axis of X relatively to the former curve by a distance $\dfrac{\pi}{2}$ corresponding to a quarter of a cycle or vibration period.

Apart from the demonstration it is easily seen that the maximum velocity of the vibrating particle corresponds to displacement $= 0$, while the maximum displacement corresponds to velocity $= 0$.

The equation $y = r \sin \omega t$ assumes that for $y = 0$, $t = 0$, i. e., the particle is assumed to be at the middle of its path at the beginning of the time interval considered. Hence the curve representing the motion passes through the origin. If when $t = 0$ the particle occupies any other position, the phase being for example an angle δ in advance of the zero position, the equation may be written $y = r \sin (\omega t + \delta)$ or $y = r \sin (\omega x + \delta)$, which is a more general form than the preceding. δ is called the *angle of epoch*.

Fig. 24 illustrates the construction of a curve representing a simple harmonic vibration. The abscissas are made proportional to times, and hence to arcs on the reference circle, i. e., to the angle θ. The ordinates are made proportional to displacements parallel to Y, i. e., to $\sin \theta$.

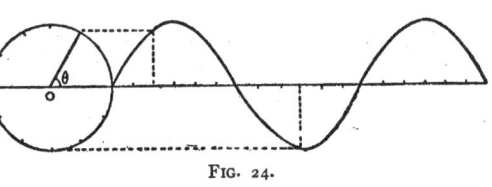

FIG. 24.

In the study of vibrations it is often more convenient to measure the phase-angle from the axis of Y rather than from X, and to use cosine rather than sine functions. In this case a vibration parallel to Y is represented by the equation $y = r \cos (\omega t + \delta)$ and the corresponding harmonic curve by the equation $y = r \cos (\omega x + \delta)$.

102. Permanent Record of Vibrations.—Trace by pencil carried at end of elastic vibrating bar against sheet of paper, moved uniformly at right angles to direction of vibration. Curves drawn on smoked glass by style attached to uniformly-moved vibrating tuning-fork, or on smoked glass carried by vibrating tuning-fork, the style being moved uniformly in a straight line.

103. Equation of Wave Motion in Elastic Medium. — From the equation of the harmonic curve we can derive the equation of a simple wave propagated in an elastic medium, as, for example, the sound-wave.

From the principles of wave motion it follows that the instantaneous value of y at time t for any point of the wave whose abscissa is x is the same as the value which it had at $x = 0$ (the source of the wave disturbance) at a previous time earlier than that under consideration by the interval which has elapsed between the starting of the wave from $x = 0$ and its arrival at x.

Denoting by T, as before, the time of one complete vibration and by m the number of cycles (in general not an integer) that have elapsed since the wave started from the origin, this interval $= mT$. Hence

$$y = r \sin \omega (t - mT) = r \sin \frac{2\pi}{T} (t - mT) = r \sin 2\pi \left(\frac{t}{T} - m\right).$$

If we denote by λ the length of the wave under consideration, $m\lambda = x$

and $m = \dfrac{x}{\lambda}$, since m waves have been generated while the disturbance was

travelling from $x = 0$ to x. Hence $y = r \sin 2\pi \left(\dfrac{t}{T} - \dfrac{x}{\lambda}\right)$, which is

the form in which the equation of the elastic wave is usually given.

As the velocity of propagation of the wave is $v = n\lambda = \dfrac{\lambda}{T}$, ($n$ = fre-

quency) this equation may also be written in the form

$$y = r \sin 2\pi \left(\frac{t}{T} - \frac{x}{vT}\right).$$

It will be seen on consideration that the equation of the wave is that of the curve shown in Fig. 24. The axis of Y at the instant under considera-

tion, t, for which the phase is $\theta = \dfrac{2\pi t}{T}$, must pass through the point for

which the ordinate is $y = r \sin 2\pi\dfrac{t}{T}$. This defines the location on X of

$x = 0$ at that instant.

104. Combination of Parallel Vibrations and of Harmonic Curves. — It is evident that for each value of x on the resultant curve the ordinate will be equal to the algebraic sum of the ordinates of the component curves. Hence if the equations of these are $y_1 = r_1 \sin (\omega_1 x + \delta_1)$ and $y_2 = r_2 \sin (\omega_2 x + \delta_2)$ the equation of the resultant curve will be $y = r_1 \sin (\omega_1 x + \delta_1) + r_2 \sin (\omega_2 x + \delta_2)$.

This equation represents a periodic curve, since if $\omega_1 x$, $\omega_2 x$ both be increased by 2π or $n2\pi$, the values of y recur. The period of the resultant curve depends upon the ratio of ω_1 to ω_2. If this reduced to its lowest term is $p : q$ the resultant period will be the time taken to make p vibrations of the first or q vibrations of the second component.

In general the nature of the results is better apprehended from a study of the graphical combination of such curves than from analysis.

Various methods have been described for drawing such compound harmonic curves mechanically.

105. Applications.—Curves drawn with tuning-forks, the smoked plate carried by one fork, the style by another. Beat Curves.

Mechanical synthesis of harmonic curves. Curve-tracing machines of Donkin and Cady. Kelvin's Tide-Predicting Machine. Synthesis and analysis by Harmonic Analyzer of Michelson.

106. Composition of Parallel Vibrations of Same Period.— A case of the composition of simple harmonic curves which is of great importance because of its application in theoretical optics is that in which the periods of the components are the same. Under these circumstances the resultant is a simple harmonic curve having the same period as that of the components. This will appear from the following consideration.

Writing $\omega_2 = \omega_1$ the equation of the resultant curve becomes

$$y = r_1 \sin (\omega_1 x + \delta_1) + r_2 \sin (\omega_1 x + \delta_2)$$

whence by expanding the values of $\sin (\omega_1 x + \delta_1)$, $\sin (\omega_1 x + \delta_2)$ we derive the equation

$$y = (r_1 \cos \delta_1 + r_2 \cos \delta_2) \sin \omega_1 x + (r_1 \sin \delta_1 + r_2 \sin \delta_2) \cos \omega_1 x.$$

In this equation we may write $r_3 \cos \delta_3 = r_1 \cos \delta_1 + r_2 \cos \delta_2$ (1) and $r_3 \sin \delta_3 = r_1 \sin \delta_1 + r_2 \sin \delta_2$ (2), subject to a subsequent determination of the values of r_3 and δ_3. Then

$$y = r_3 \cos \delta_3 \sin \omega_1 x + r_3 \sin \delta_3 \cos \omega_1 x = r_3 (\cos \delta_3 \sin \omega_1 x + \sin \delta_3 \cos \omega_1 x)$$

or $y = r_3 \sin (\omega_1 x + \delta_3)$

This represents a simple harmonic curve of period ω_1, amplitude r_3, and epoch δ_3.

The values of r_3 and δ_3 are easily found as follows:—

By squaring equations (1), (2), adding the results and simplifying we have $r_3{}^2 = r_1{}^2 + r_2{}^2 + 2r_1 r_2 \cos (\delta_1 - \delta_2)$, whence the amplitude of the resultant is

$$r_3 = \sqrt{r_1{}^2 + r_2{}^2 + 2r_1 r_2 \cos (\delta_1 - \delta_2)}.$$

Dividing (2) by (1) we have $\tan \delta_3 = \dfrac{r_1 \sin \delta_1 + r_2 \sin \delta_2}{r_1 \cos \delta_1 + r_2 \cos \delta_2}.$

The cases where the amplitudes r_1, r_2 are equal should be particularly noticed. If in addition to this equality $\delta_1 - \delta_2 = 0, r_3 = 2r_1$; that is, if the components are in the same phase the amplitude of the resultant curve is double that of either component. If $\delta_1 - \delta_2 = \pi, r_3 = 0$; that is, if the components are in precisely opposite phases, the amplitude of the resultant = 0 and the curve is reduced to a right line coinciding with the axis of X. For intermediate values of the phase-difference the amplitude varies between these two extremes.

It follows furthermore from what precedes that the addition of any number of simple harmonic curves of the same period gives as a resultant a simple harmonic curve also of the same period. For the components may be combined successively, each with the resultant of those previously taken.

107. Combination of Rectangular Harmonic Vibrations.— Measuring the phase-angles from Y and X respectively, the component vibration parallel to Y may be represented by the equation $y = r \cos \omega t$, that parallel to X by the equation $x = r_1 \cos (\omega_1 t + \delta)$.

For purposes of illustration it will suffice to consider the simplest case, that in which the periods are the same and hence $\omega_1 = \omega$. We have, therefore, $x = r_1 \cos \partial \cos \omega t - r_1 \sin \partial \sin \omega t$. But $\cos \omega t = \dfrac{y}{r}$, $\sin \omega t = \sqrt{1 - \dfrac{y^2}{r^2}}$. Hence $x = r_1 \cos \partial \left(\dfrac{y}{r} \right) - r_1 \sin \partial \left(\sqrt{1 - \dfrac{y^2}{r^2}} \right)$, and

$$x^2 + r_1{}^2 \cos^2 \partial \left(\frac{y^2}{r^2} \right) - 2\, r_1 x \cos \partial \left(\frac{y}{r} \right) = r_1{}^2 \sin^2 \partial - r_1{}^2 \sin^2 \partial \left(\frac{y^2}{r^2} \right),$$

which may be written

$$\frac{x^2}{r_1{}^2} + \frac{y^2}{r^2} - \frac{2\,x\,y}{r_1 r} \cos \partial = \sin^2 \partial.$$

This is the equation of an ellipse.

If $\partial = 0$, $\left(\dfrac{x}{r_1} - \dfrac{y}{r} \right)^2 = 0$ and $\dfrac{x}{r_1} = \dfrac{y}{r}$,

which is the equation of a straight line passing through the origin and making an angle $\tan^{-1} \left(\dfrac{r}{r_1} \right)$ with X.

If $\partial = \frac{1}{4}\,(2\pi)$, $\dfrac{x^2}{r_1{}^2} + \dfrac{y^2}{r^2} = 1$, the equation of an ellipse with its axes lying in X and Y. If in addition $r = r_1$, $x^2 + y^2 = r^2$, and the ellipse becomes a circle.

If $\partial = \frac{1}{2}\,(2\pi)$, $\dfrac{x}{r_1} = -\dfrac{y}{r}$, the equation of a straight line through the origin and making an angle $\tan^{-1} \left(-\dfrac{r}{r_1} \right)$ with X.

If $\partial = \frac{3}{4}\,(2\pi)$, $\dfrac{x^2}{r_1{}^2} + \dfrac{y^2}{r^2} = 1$, and if also $r = r_1$, $x^2 + y^2 = r^2$.

The ellipses represented by the preceding equations are always inscribed in a rectangle of which $2r$, $2r_1$ are the sides. The points of contact of the ellipse with the sides of this rectangle may be found by determining the intersections of the curve with the lines $x = \pm r$, $y = \pm r_1$.

If the periods of the vibrations to be combined have any other ratio, e. g., $1 : 2$, or $1 : n$, in the general equation we must make $\omega = 2\omega_1$ or $\omega = n\omega_1$. But except in a few special cases the resulting curves are of very complex character and are more readily studied by graphical methods. (See text-book.)

If the difference of phase represented by ∂ varies gradually from 0 to 2π the curve will gradually pass in succession through all its corresponding variations. Fig. 26 illustrates these for the several ratios there indicated.

108. **Applications.**—Blackburn's Pendulum (Fig. 25); Harmonograph Pendulums; Wheatstone's Kaleidophone; Lissajous' Curves (Fig. 26); Lissajous' Comparator. Mechanical Combination by Machines of Pickering, Donkin, and others.

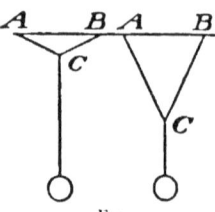

Fig. 25.

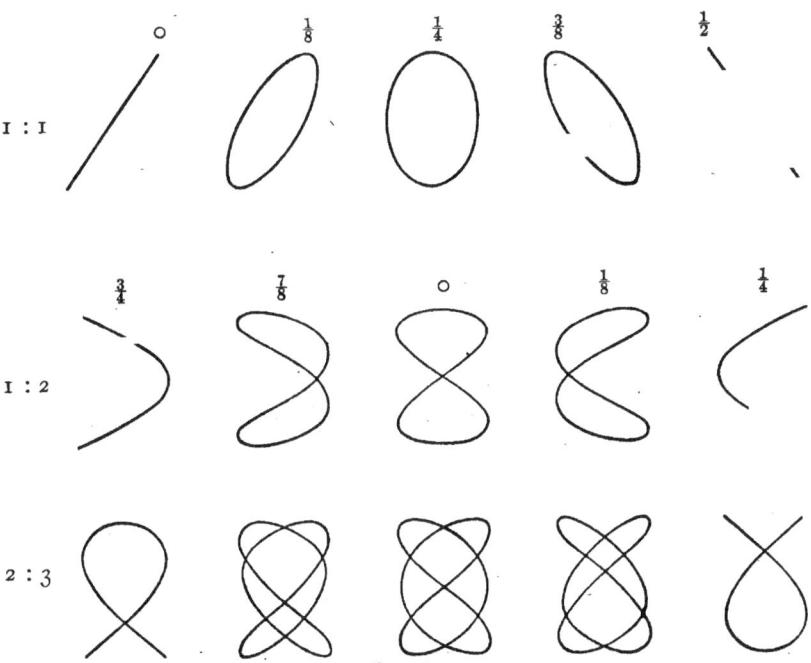

Fig. 26.

109. Fourier's Theorem.—"If any arbitrary periodic curve be drawn, having a given wave-length λ, the same curve may always be produced by compounding harmonic curves (in general infinite in number, having the same axis, and having λ, $\frac{1}{2}\lambda$, $\frac{1}{3}\lambda$, . . . for their wave-lengths.

"The only limitations to the irregularity of the arbitrary curve are, first, that the ordinate must be always finite; and, secondly, that the projection on the axis of a point moving so as to describe the curve must move always in the same direction.

"These conditions being satisfied, a wave of the curve may have any form whatever, including any number of straight portions.

"Analytically the theorem may be expressed as follows:—

"It is possible to determine the constants C, C_1, C_2, etc., a_1, a_2, etc., so that a wave of the periodic curve defined by the equation

$$y = C + C_1 \sin\left(\frac{2\pi x}{\lambda} + a_1\right) + C_2 \sin\left(2\frac{2\pi x}{\lambda} + a_2\right) + \cdots$$

or
$$y = C + \sum_{i=1}^{i=\infty} C_i \sin\left(\frac{2i\pi x}{\lambda} + a_i\right)$$

shall have any proposed form, subject to the conditions mentioned above."

The preceding statement is extracted from Donkin's "Acoustics" (Clarendon Press Series).

110. Vibrations of Elastic Bodies.—Since in the case of an elastic body subjected to any kind of deformation the restoring force within certain limits is proportional to the displacement, it follows that vibrations of small amplitude performed under the influence of elasticity are simple harmonic in character, and isochronous.

In case the amplitude of the vibration is so great that the restoring force is not at each point proportional to the first power of the displacement, the motion while still periodic is no longer simple harmonic in its character. It can be shown by analysis that under these circumstances the vibration is a compound harmonic one, represented by the resultant of a series of simple harmonic vibrations of frequencies n, $2n$, $3n$, etc., the particular harmonics present being determined by the law of variation of the restoring force.

Experimental evidence of these effects is clearly seen in the case of a vibrating tuning-fork. When gently bowed only the fundamental (frequency $= n$) can be heard, but with strong bowing the octave (frequency $= 2n$), the fifth of the octave (frequency $= 3n$), and even higher harmonics also appear.

A like occurrence, to be explained in a similar manner, is the occasional appearance of harmonic notes in telephonic transmission when a microphone transmitter is used.

The mathematical principles underlying these several phenomena are stated in § 109, p. 49.

111. Acoustic Vibrations.—The formula $T = 2\pi\sqrt{\dfrac{y}{a}} = 2\pi\sqrt{\dfrac{d}{a}}$ (in which d is the displacement) enables us to obtain the laws of vibration of bodies vibrating under the influence of elasticity.

The following demonstrations are of importance in acoustics, as giving the laws of vibration of sounding bodies.

Let $n = \dfrac{1}{T}$ be the frequency of vibration. Then $n = \dfrac{1}{2\pi}\sqrt{\dfrac{a}{d}} = \dfrac{1}{2\pi}\sqrt{\dfrac{F}{Md}}$, where F is the restoring force and M the mass in vibration.

For a bar or wire, $M = $ length $\times$ cross-section $\times$ density $= l\,s\,w$, and therefore $n \propto \sqrt{\dfrac{F}{lswd}}$.

For a bar or wire executing longitudinal vibrations, $F = \dfrac{Esd}{l}$, E being Young's modulus of elasticity; whence $n \propto \sqrt{\dfrac{Esd}{l^2swd}} \propto \dfrac{1}{l}\sqrt{\dfrac{E}{w}}$.

Hence for longitudinal vibrations the frequency is independent of the section of the bar, and varies inversely as the length, inversely as the square root of the density and directly as the square root of the modulus of elasticity of its material.

For an elastic string vibrating transversely, calling P the tension of the string when deflected, $F : P : : 2d : \tfrac{1}{2}l$. (See Fig. 27.)

Hence $F = \dfrac{4Pd}{l}$ and $n \propto \sqrt{\dfrac{4Pd}{l^2 swd}} \propto \dfrac{1}{l}\sqrt{\dfrac{P}{sw}} \propto \dfrac{1}{lr}\sqrt{\dfrac{P}{w}}$,

as $s = \pi r^2$.

Hence for such a vibrating string the frequency varies inversely as the length of the string, directly as the square root of the tension, inversely as the diameter and inversely as the square root of the density.

For a vibrating elastic rod of length l, breadth b, depth k, $F = \dfrac{Abk^3 d}{l^3}$,

A being a constant. Hence $n \propto \sqrt{\dfrac{Abk^3 d}{(lswd)l^3}} \propto \sqrt{\dfrac{bk^3}{(lbkw)l^3}} \propto \dfrac{k}{l^2}\sqrt{\dfrac{1}{w}}$.

That is, for a bar vibrating transversely the frequency is independent of the breadth, inversely as the square of the length, directly as the depth, and inversely as the square root of the density.

These laws of variation of frequency are independent of the particular manner in which the bar is supported, only the absolute frequency being affected by this.

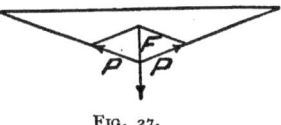

FIG. 27.

112. Circular Pendulum.—The time of vibration of a circular pendulum swinging in an indefinitely small arc may be found as follows:—The vibration is simple harmonic since the accelerating component is $F = W \sin \theta = W\theta$ for small angles. (Fig. 28.) Let $OP = l$ and $PN = r$.

Then $T = 2t = 2\pi \sqrt{\dfrac{r}{a}} = 2\pi \sqrt{\dfrac{l \sin \theta}{g \sin \theta}} =$

$2\pi \sqrt{\dfrac{l}{g}}$, and $t = \pi \sqrt{\dfrac{l}{g}}$.

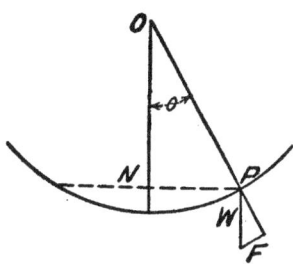

FIG. 28.

113. Cycloidal Pendulum.—The law of curvature of the cycloid is such that with the cycloidal pendulum the accelerating (tangential) component of the weight of the vibrating particle ($W \sin \theta$) is always proportional to the arc of displacement. Hence the vibrations are isochronous for all arcs.

This property of the cycloid follows directly from the fact that with it the length of the arc included between the vertex and any point E (Fig. 29) is

$d = 4a \cos \dfrac{\phi}{2}$ where a is the radius of the gen-

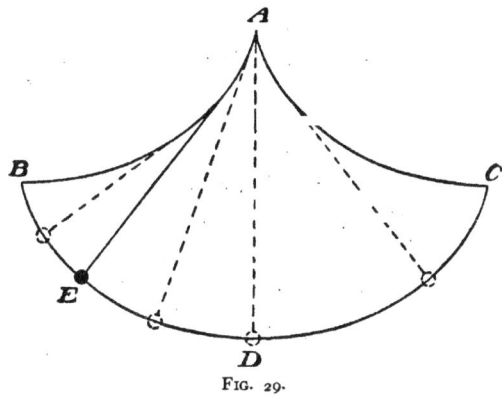

FIG. 29.

erating circle and ϕ the angle made with the vertical by this radius drawn through E. Calling θ the angle of deviation when the vibrating particle

is at E, $\dfrac{\phi}{2} = (90° - \theta)$. The restoring force at E is $F = W \sin \theta =$

$W \cos (90° - \theta) = W \cos \dfrac{\phi}{2}$. Hence $F \propto \cos \dfrac{\phi}{2} \propto d$.

It will also be seen from the demonstration in § 112 that the time of

vibration for all arcs is $t = \pi \sqrt{\dfrac{l}{g}}$.

114. Damped Vibrations.—When a vibrating body executes free vibrations the amplitude of these gradually diminishes owing to the energy expended in overcoming external or internal resistances. This effect is called *damping*.

In general the effect of the resistance is proportional to the velocity simply, and the diminution of amplitude is proportional to the amplitude itself at each instant. That is, denoting amplitudes by y and times by x,

$\dfrac{dy}{dx} = - ky$ whence $\dfrac{dy}{y} = - k\,dx$. Hence by integration

$$\log y = - kx + constant \text{ and } y = ce^{-kx}.$$

It follows from this equation that if we designate by y_m and y_n respectively the amplitudes of the decreasing vibration after corresponding times x_m, x_n, and if during the interval $x_n - x_m$, N vibrations have been made, then

$\dfrac{\log y_m - \log y_n}{N} = \Lambda$, a constant, which is known as the *logarithmic decrement*

of the vibration, and is evidently the decrement per vibration of the logarithm of the amplitude.

It will also be seen that the vibration of a particle subject to damping may be represented graphically by replacing the reference circle in Fig. 24 by a logarithmic spiral.

The periodic curve representing a damped vibration must be characterized by a diminution of its ordinates from those of the sine curve representing the corresponding displacement with an undamped vibration of the same period, which follows the law represented by the equations just demonstrated. Hence the equation of such a harmonic curve will take the form $y = e^{-kx} r \sin (\omega x + \delta)$.

Fig. 30 illustrates the method of constructing the curve representing a damped simple harmonic vibration. The abscissas are made proportional to times and hence to the angle θ. The ordinates are made proportional to the value for the angle θ of the

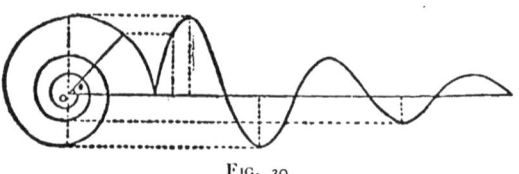

Fig. 30.

ordinate of the extremity of the radius vector of the logarithmic spiral of reference, *i. e.*, to $r_\theta \sin \theta$.

115. Stroboscopic Method of Studying Periodic Motion. — When a rapidly moving body is illuminated by a flash of light of such brief duration that the body does not traverse a perceptible space during its continuance, the object appears to be at rest. This is the case, for example, with a swiftly revolving wheel illuminated by the electric spark. If there is a series of brief illuminations the object is seen in a corresponding series of positions. If a body possessing a periodic motion is illuminated by a series of recurrent flashes, and if the interval between each flash is such as to coincide exactly with that period (or with n times the period) the body will appear to be absolutely at rest, however great its velocity may be. It will be seen continuously owing to the persistence of vision. If the interval between the flashes is slightly less than the periodic time the body will apparently possess a slow retrograde motion. If the interval is slightly greater than the periodic time, there will be an apparent slow forward motion of the body. In this way changes of configuration or of form may be studied.

The flashes may be obtained by any convenient method, as by the electric spark or by throwing the light from an electric arc through a rotating disc with radial slits; or the moving object while continuously illuminated may be viewed directly by the eye, looking through a rotating slotted disc.

116. Illustrations and Applications. — (a) Single flash. Flashlight photography. Rapid-shutter photography. Study of trotting horses, etc., by Muybridge. Spark-photography of flying bullets by Mach and Boys.

(b) Stroboscope.—Determination of period of rotating wheel. Study of motion of chain of small bones in human ear by Mach. Determination of frequency of tuning-fork. Stroboscopic study of alternating arc in Rogers Laboratory, M. I. T. Use of microstroboscope in telephonic studies in Rogers Laboratory.

Toys; e. g., Phenakisticope of Plateau. Stroboscopic discs of Stampfer. Dedaleum of Horner.

(c) Kinetoscope; Kinematograph. Early application of photography by Muybridge to study of gait and postures of animals. Representation of movement of express train, of mountain ascents, etc. Representation of progressive motion of waves by Müller and Wood. Study of plant growth.

117. Free and Forced Vibrations.—If a body so conditioned as to be capable of entering into a state of vibration is subjected to the action of a disturbing impulse and then left free, it will execute what are called *free vibrations*, the period of these being determined by the mass, form, and mechanical conditions of the body. If such a body is subjected to the continued action of a periodically varying force, external to itself, this will act to produce in the body vibrations of the same period as that of the force variation. These are called *forced vibrations*.

If the varying force coincides in period with any vibration which it is possible for the body to execute, extremely strong forced vibrations may be produced, even though the effect of each separate periodic action of the force is small. In this case vibrations thus excited are often called *sympathetic vibrations* or *syntonic vibrations*.

When the action of the force is reciprocating, so that periodic reversals in its direction occur, it can only produce strong sympathetic vibrations in a body which is capable of vibrating in the same period. But if the force acts intermittently, producing a series of equally timed impulses, it may excite sympathetic vibrations in a body which is capable of vibration in any period which is an integral number of times that of the impulses.

Illustrations of the former of these conditions are found in the acoustic phenomena of the sympathetic tuning-forks and resonance.

The latter conditions are found in many cases of forced mechanical vibrations, as in the ringing of a church bell, in Galileo's experiment in which a heavy pendulum is set into vibration by properly timed puffs of the breath, in the vibrations of suspension and other bridges excited by marching bodies of troops, and in the vibration of mill buildings by machinery.

118. Applications.—Hardy's · "Noddy" used in pendulum experiments. Frahm's Resonance Apparatus. Hartmann and Braun's Frequency Meter for Alternating Currents. Helmholtz's Harmonic Tuning-forks. Clocks governed by action of electro-magnet upon Pendulum.

119. Phase Relation.—There is always a certain difference of phase between the force which acts to produce the forced vibration and the resulting forced vibration itself, the latter lagging behind the former. The amount of this difference depends upon the relation between the periods of the force and the free vibration. If these are precisely the same, the case of perfect unison, the phase-difference can be shown to be $\frac{1}{2}\pi$, that is, one-fourth of a period. If the natural frequency is the greater, the lag of the forced vibration is less than a quarter period; if the natural frequency is the lesser, the lag is between one-quarter and one-half period.

ILLUSTRATION: Phenomena observed with Sympathetic Pendulums.

120. Coexistence of Free and Forced Vibrations.—When a body is set into forced vibration by the action upon it of a periodic force there is often noticed at the beginning of the operation, before the forced vibration is fully established, a rhythmic variation in the amplitude of the vibration actually performed. In the case of an acoustic vibration "beats" are heard.

This phenomenon is caused by the simultaneous existence of a free vibration, due directly to the disturbance of the vibratory body and the forced vibration impressed upon the body. If these two periods are not exactly alike beating must occur.

ILLUSTRATIONS: Beats of tuning-fork electrically driven by a governing fork of same frequency. Beats of electrically excited musical string. Frahm's Resonance Top.

121. **Resonance.**—In acoustics the term resonance is usually applied to the phenomena of forced vibrations, especially when they are produced in a mass of air.

The small density of air enables it to enter into a state of forced vibration very readily even when the period of this is considerably removed from the period of free vibration, although the resonance rises rapidly in amount as the period of the forced vibration approaches that of the free vibration of the mass of air.

Any mass of air or other body which is capable of entering into a state of vibration will respond by resonance to sounds having the same pitch (and hence the same vibration-frequency) as any which it can emit in consequence of its own free vibration.

ILLUSTRATIONS OF RESONANCE OF MASS OF AIR: Tuned Bottles. Helmholtz's resonators. Open and stopped organ pipes.

In case a body is capable of vibration in several different periods it will respond to any one of the several notes of corresponding vibration-frequency. It will also respond to several such sounds simultaneously.

ILLUSTRATIONS: Multiple resonance of organ pipes. Simultaneous resonance of organ pipe to two or more forks.

122. **Conditions affecting Damping.** — The more readily a body enters into a state of forced vibration, the more readily are its vibrations damped. For example, a mass of air in a resonator can be forced by a powerful tuning-fork to vibrate in a period far removed from its natural rate of free vibration; but the sound of such a vibrating mass of air persists only for a very brief period. On the other hand, a tuning-fork can only be made to execute forced vibrations when the exciting fork is very closely in unison with it; but vibrations excited in any manner in the fork are very persistent.

Whenever a vibrating body causes sympathetic vibrations in a second body, the former loses its own energy of vibration more rapidly in proportion to the strength of the resonance which it causes. Hence the damping of the vibrations of a body in proximity to others capable of vibrating at the same rate as itself will be more rapid than if it is removed from the neighborhood of such bodies. The reaction of the sympathetically vibrating body upon the other is always such as to oppose its motion.

INTERFERENCE OF WAVES.

123. **Phenomena.**—If two trains of parallel waves of equal length meet one another and coalesce, the waves constituting the resultant train will possess an amplitude equal to the algebraic sum of the amplitudes of the components. The wave length will remain unchanged. If the waves

are in similar phase the resultant wave will be of greater amplitude than either component, if they are in unlike phase, less. In the particular case where the amplitudes of the coalescing waves are equal and their phase opposite they will neutralize each other and the amplitude of the resultant wave will be zero. This effect is known as *interference*. If the amplitudes of the waves thus compounded are unequal, the interference will be only partial.

This phenomenon is considered analytically in § 106, p. 47 of these Notes.

In most cases of interference the interfering trains of waves start from the same source and reach the place of coalescence by two different paths, one of which is half a wave length or an odd number of half wave lengths longer than the other. Under these circumstances the two sets of waves always meet trough to crest and so interfere, completely or partially, according to their relative amplitudes.

In the case of sound waves meeting in opposite phase, *i. e.*, condensation to rarefaction, the addition of one sound to another sound of the same pitch and loudness, and hence of the same wave length and amplitude, may produce silence. In like manner the addition of light to light may produce darkness.

ILLUSTRATIONS: Water Waves.—Absence of tide at Batsha, where the tidal wave reaches the port by two channels of different length. Increased tide at certain points on eastern coast of England arising from interference of wave trains travelling respectively northward from English Channel and southward from sea above Scotland.

Waves in cords.—Stationary waves in Melde's experiment.

Sound waves.—Herschel's trombone apparatus. Stationary waves formed in free air by coalescence of trains of waves, respectively direct and reflected from a wall.

Light waves.—Colors of thin films, as in case of soap bubble.

GRAVITATION.

124. **Law of Universal Gravitation.** — *Every particle of matter in the universe attracts every other particle with a force varying directly as the product of their masses, and inversely as the square of their distance.*

Hence the mutual gravitation of two masses m, m', at a distance d, is

$$G = \phi \frac{mm'}{d^2}.$$

The general theory of gravitation, as proved by Newton during the years 1666 to 1687, followed from his establishment of the following facts:—

"1. That the force by which the *different* planets are attracted to the sun is in the inverse proportion to the squares of their distances. [1. Results from 3d Law of Kepler.]

"2. That the force by which the *same* planet is attracted to the sun, in different parts of its orbit, is also in the inverse proportion to the square of the distance. [2. Results from 1st and 2d Laws of Kepler.]

"3. That the earth also exerts a force on the *moon*, and this force is identical with the force of *gravity*.

"4. That bodies thus act on *other* bodies besides those which revolve around them; thus the sun exerts such a force on the moon and satellites, and the planets exert such forces on *one another*.

"5. That this force thus exerted by the general masses of the sun, earth, and planets arises from the attraction *of each particle* of these masses; which attraction follows the above law, and belongs to all matter alike."
—WHEWELL.

The Laws of Kepler referred to above were determined by that astronomer wholly from observation (1609; 1619). They are as follows:—

1. The planets all revolve in ellipses with the sun in one focus.

2. The radius vector describes areas proportional to the times.

3. The squares of the periodic times of the planets are proportional to the cubes of their mean distances from the sun.

125. Value of Gravitation Constant. — The constant ϕ in the formula for gravitation denotes the attraction existing between two masses of one gram each at a distance of one centimeter apart. Its value has been determined by the method of Cavendish (1798) in which a small metallic sphere carried on the arm of a torsion balance is attracted by a heavy lead sphere. The force between the two spheres is known from the torsion developed in the wire of the balance. Their masses and distance apart are also known, so that the value of ϕ can easily be computed. Results by Boys, who used a quartz fibre suspension (1895) give as its value $\phi = 6.6576 \times 10^{-8}$. By comparing the attraction of the lead sphere on the sphere of the torsion balance with the attraction of the earth on the same, the mean density of the earth can be determined. The value corresponding to the value of ϕ given above is $\varDelta = 5.5270$.

Values of ϕ agreeing closely with that given above have also been obtained by other methods. Thus Poynting (1891) measured the attraction between two lead spheres by means of a chemical balance of great sensitiveness.

The law of gravitation appears to hold throughout all distances, varying from interplanetary spaces certainly to within a few centimeters and presumably until the attracting bodies are in contact. ϕ is independent of the material of the masses, and of the medium separating them. Also gravitation is in no case directive or polar in character.

126. Diminution of Gravity without Surface of Sphere. — Newton showed that the resultant effect of all the particles of matter composing a homogeneous sphere upon a particle outside of it is the same as if the total mass of the sphere were concentrated at its center. Hence, calling G, G_h, the respective attractions upon a body at the surface of a sphere of radius R, and at a distance h above the surface,

$$G : G_h : : \frac{1}{R^2} : \frac{1}{(R+h)^2};$$

whence

$$G_h = G \frac{R^2}{(R+h)^2},$$

or

$$G_h = G \left(1 - \frac{2h}{R} \right), \text{ approximately.}$$

The preceding formula can be applied to the diminution of gravity on ascending above the surface of the earth, where G, G_h are the weights of the body. Also, it follows that

$$g_h = g \left(1 - \frac{2h}{R} \right) \text{ approximately.}$$

VARIATION OF g WITH ELEVATION.

	Elevation ft.	g cm.	
San Francisco	375	979.951	Mendenhall
Lick Observatory, Mt. Hamilton	4205	979.646	"
Denver	5374	979.595	Putnam
Pike's Peak	14085	978.940	"
Kawaihae	8	978.798	Preston
Kalaieha	6660	978.485	"
Waiau, Summit Mauna Kea	13060	978.055	"
Do. Reduced to Lat. of Kawaihae		978.067	"
Tokio	0	979.84	Mendenhall
Fuji, Summit	12441	978.86	"
Do. Reduced to Lat. Tokio		978.65	"

Von Jolly (1881) made a direct determination of the diminution of g in ascending, by the use of an equal-arm balance with two sets of pans, one close to the beam, the other at a distance below it, and consequently nearer the surface of the earth. With a distance of 121 metres between the pans, a mass of 5 kilograms weighed 32 milligrams more when in the lower than when in the upper pan.

Von Jolly furthermore found that by placing a sphere of lead of mass

·5775 kilograms immediately under the movable mass when thus lowered, the weight of the latter was increased by one-half of a milligram. This was due to the gravitational attraction between the two spheres. It follows directly from this that the earth's attraction on the movable mass was 10 million times that of the lead sphere, and hence that the mean density of the earth should be about 5.7.

Richarz and Krigar-Menzel using a modification of this method obtained a value of $\varDelta = 5.5$, agreeing very closely with that obtained by Boys by the torsion balance.

127. Instruments for Direct Measure of Variations in Gravity.— Mass supported by spring; Siemens' bathometer. Threlfall and Pollock's quartz-thread gravity balance. Von Sterneck's barymeter; oblique column of mercury balanced on knife-edge support.

128. Effect of Spheroidal Form of Earth.—Analysis and observation both show that the force of gravity increases from the equator to the poles. The actual gain of weight of a body carried from the former to the latter latitude would be about $\frac{1}{189}$ of its original value. Of this approximately $\frac{1}{288}$ arises from the centrifugal force caused by the rotation of the earth. The remainder, about $\frac{1}{545}$, is due to the fact that the earth is an oblate spheroid.

$$\text{At the poles} \quad \mathring{g} = 983.194 \ cm.$$
$$\text{In latitude } 45° \ g = 980.630 \ cm.$$
$$\text{At the equator } \mathring{g} = 978.066 \ cm.$$

VALUE OF g_0 FOR ANY PLACE.—If g_0 be the value of the acceleration of gravity in latitude o° and at sea-level, the acceleration g_λ in latitude λ and at altitude h will be $g_\lambda = g_0 \left(1 - 0.005243 \sin^2\lambda\right) \left(1 - \frac{2h}{R}\right)$.

This is the value which g_λ would have in mid-air. In the actual case of stations in mountainous regions various particular formulæ have been used, as no general formula can be deduced which will take all local disturbances into account. (See § 150, p. 68, for a further consideration of the subject.)

129. Gravity within Sphere. — It can be shown that within a homogeneous sphere the weight of a mass would be directly proportional to its distance from the center. But this condition is very far from being true of the earth owing to the increased density of deeper strata. In fact, g at first increases with increasing depth. Below a certain but unknown depth it must obviously decrease.

130. Falling Bodies.—Laws.

I. *The acceleration produced by gravity is independent of the mass;* whence *the velocity of a freely falling body is independent of its mass.*

EXPERIMENTAL PROOF.—Cannon-balls of different masses dropped from tower. (Galileo, about 1590, at Pisa.)

THEORETICAL PROOF.—Calling M, M' the masses of two bodies, G, G' their attraction to the earth, a, a' the accelerations produced, we have

$$G : G' :: M : M',$$
but $\qquad Ma : M'a' :: G : G';$
whence $\qquad a = a'.$

II. *The acceleration is independent of the material of the body.*

EXPERIMENTAL PROOFS.—(a) Falling balls of metal and cork; (b) guinea and feather tube; (c) proofs of Newton and Bessel by pendulum, to be discussed later.

III. *Velocity is proportional to the time of descent.*
IV. *Velocity is proportional to the square root of distance fallen through.*
V. *Space described is proportional to the square of the time.*

III., IV., V., follow from the fact that the motion is uniformly accelerated.

131. Verification of Laws by Experiment.—The three last-mentioned laws of falling bodies may be investigated experimentally by means of various devices invented for the study of uniformly variable motion, including that produced by gravity. Of these the following are particularly worthy of mention:—

1. *Inclined Plane* (Galileo). Apply preceding formulæ. (See § 26, p. 8.)

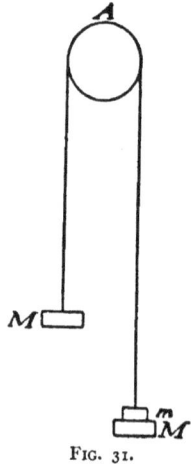

FIG. 31.

2. *Atwood's Machine* (1780) (Fig. 31). The equal large masses, M, M are connected by a flexible cord running over a pulley A. The addition of a small mass m produces a uniformly accelerated motion, in which the acceleration is found from the equation

$$w = mg = a\,(2\,M + m), \text{ whence } a = g\,\frac{m}{2M + m}.$$

This value of a being substituted in the general formulæ, we have the velocity acquired by the mass and the space traversed by it, which may be compared with the results of experiment.

The effect of the pulley A must also be allowed for. This may be computed from its moment of inertia, but it is best determined by experiment. It will be the same as if the masses M, M were increased by a constant amount M'. Making this correction, $a = g\,\dfrac{m}{2M + M' + m}.$

The following demonstration of the value of a with Atwood's machine is based on the principles of Energy.

If the heavier mass descends through a distance h and the system acquires thereby a velocity v, then $wh = mgh = \frac{1}{2}(2M + m)v^2 = \frac{1}{2}(2M + m)(2ah)$, observing that $v^2 = 2ah$, since the motion is uniformly accelerated.

Solving this equation relatively to a, we have $a = g \dfrac{m}{2M + m}$, as before.

3. *Barbouze's Machine*. In this machine, a lamp-blacked cylinder is attached to the axis of the pulley of an Atwood's machine. Against this rubs a style attached to a tuning-fork, which marks equal intervals of time by its vibrations, and from the relative length of the sinuosities produced with increasing velocity, the laws can be determined experimentally.

A simpler method, used at the Institute in 1869, consists in causing a freely falling glass plate, covered with lamp-black, to press very lightly against a style carried by one of the prongs of a vibrating tuning-fork. In a later form of the apparatus the fork falls while the glass plate is fixed.

4. *Morin's Machine* (Fig. 32). A freely falling body traces its path against a cylinder moving uniformly about a vertical axis. By the combination of the two motions a parabola is traced, showing that the motion of the falling body is uniformly accelerated.

FIG. 32.

In a modified form of the machine, devised by Sir George Darwin, the pencil is fixed and the revolving cylinder falls.

5. *Additional Methods*. The laws of falling bodies are also studied by determining the velocity acquired and space traversed, by means of electric chronographs.

It is evident that any of these devices will enable us to obtain a value of g, of approximate correctness. Accurate methods of determining this will be explained in the chapter on the pendulum.

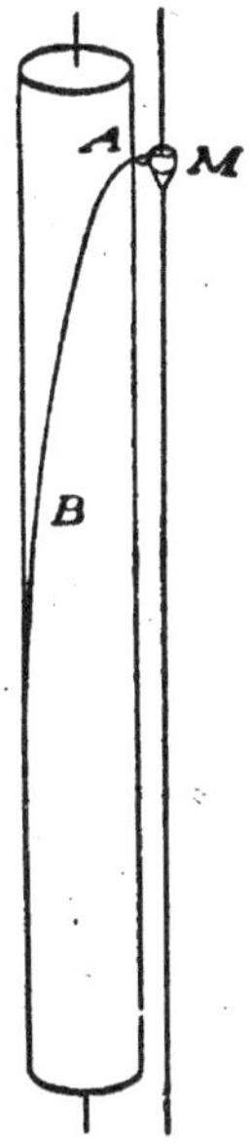

132. Case of Body falling from Great Height. — Since gravity varies inversely as the square of the distance from the center of the earth, in the case of a body falling through a great distance, this must be taken into account. For small distances the diminution may be neglected, as at the height of a kilometer it is only $\frac{1}{3188}$ of the value at the surface. It can be shown by analysis that the velocity acquired by a body falling freely to the earth from an infinite distance would be about 35,000 feet per second.

Conversely, a mass projected vertically upward with this velocity would not return to the earth.

This principle has served to disprove an early hypothesis as to the lunar volcanic origin of meteorites, and also, by an application of the Kinetic Theory to indicate the probable constitution of planetary atmospheres.

133. Projectile. — The trajectory of an unresisted projectile is a parabola.

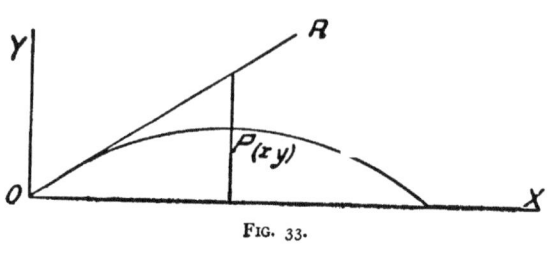

FIG. 33.

Let $O R$ (Fig. 33) be the original direction of projection and V_0 the initial velocity. Call the angle of elevation $R O X$, θ.

Let x, y, be the coordinates of any point P on the curve and t the time taken by the projectile to reach that point. Then

(1) $x = (V_0 \cos \theta) t$

(2) $y = (V_0 \sin \theta) t - \frac{1}{2} g t^2$.

By eliminating t we have

(3) $y = x \tan \theta - \dfrac{g}{2 V_0^2 \cos^2 \theta} \cdot x^2$ which is the equation

of the trajectory.

This is the equation of a parabola with a vertical axis.

From (1), (2), (3) can be found the value of the horizontal and vertical ranges, the time of flight and the elevation for maximum horizontal range.

134. Effects of Air Resistance. —The resistance of the air causes a large deviation from the parabolic trajectory and a great diminution of range in the case of rapidly moving projectiles. The maximum horizontal range is theoretically obtained when the angle of elevation is 45°, but this is practically true only for low velocities. For swift projectiles this angle is about 35°.

Other effects of air-pressure on a projectile are (*a*) gyroscopic deviation (drift) with elongated projectile from rifled gun, since the pressure tends to tilt the top of the projectile upward; (*b*) lateral deviation from the plane of projection in case of a ball rotating about a vertical axis, caused by unequal air-pressure on opposite vertical halves according as the velocity at the surface due to rotation is with or against the motion of translation; as in the case of curve pitching in baseball.

In the case of an unsymmetrical rotating projectile the air-resistance may give rise to a very complex path, as, for example, in the Australian *boomerang*, a curved club which returns to the place from which it is thrown.

135. Velocity acquired in descending Frictionless Inclined Plane. —Since $v = \sqrt{2 g s \dfrac{H}{L}}$, if $s = L$, that is, if the body traverses the whole length of the plane, $v = \sqrt{2 g H}$. This is independent of L, the length of the plane, and is equal to the velocity acquired by a freely falling body in descending through the vertical height H.

136. General Proposition. —*A body descending from a given point to a given horizontal plane will acquire the same velocity whether this descent is made vertically or obliquely over an inclined plane or over a curved surface.* This proposition is, of course, departed from in practice, because of the effect of friction and other resistances.

137. Properties of Cycloid.—The path along which a body will descend most swiftly, between two points not in the same vertical, is an inverted cycloid passing through these points with its cusp at the uppermost of them. For this reason, the cycloid is often called the *Brachystochrone.*

Another important property of the cycloid is, that the time required to descend to the lowest point of the complete inverted curve is the same, from whatever point of the curve the body may start. The cycloid is therefore a *Tautochrone.*

PENDULUM.

138. Case of Body rolling on Curve or Suspended by Flexible Cord.—See Figs. 34, 35.

139. Simple or Mathematical Pendulum. —This may be defined as *a gravitating particle suspended by a cord without weight.*

140. Time of Vibration.— For very small circular arcs

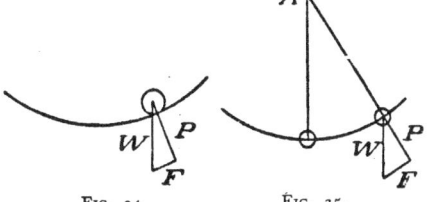

Fig. 34.   Fig. 35.

$$t = \pi \sqrt{\frac{l}{g}}. \quad \text{(See § 112, p. 51.)}$$

This is independent of amplitude. (Galileo, about 1583.)

For all circular arcs,

$$t = \pi \sqrt{\frac{l}{g}} \left[1 + \left(\frac{1}{2} \right)^2 \left(\frac{h}{2l} \right) + \left(\frac{1 \cdot 3}{2 \cdot 4} \right)^2 \left(\frac{h}{2l} \right)^2 + \left(\frac{1 \cdot 3 \cdot 5}{2 \cdot 4 \cdot 6} \right)^2 \left(\frac{h}{2l} \right)^3 + \cdots \right], \quad \text{or} \quad t = \pi \sqrt{\frac{l}{g}} \left[1 + \frac{1}{4} \cdot \frac{h}{2l} \right],$$

a very close approximation used in practice. In the formula h is the versed sine of the arc of vibration.

141. Isochronism.—If the particle moves in a cycloidal arc, the formula $t = \pi \sqrt{\frac{l}{g}}$ is true for all amplitudes. (Huygens, 1673.) (See § 113, p. 51, for proof.)

Hence, for a cycloidal pendulum, or approximately for a circular pendulum vibrating through small arcs, the vibrations are isochronous; that is, performed in equal times, independently of the amplitude.

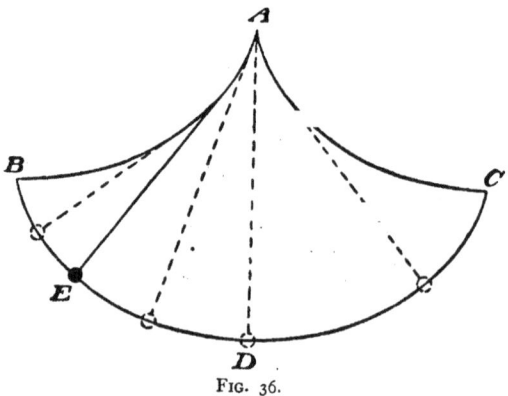

Fig. 36.

Construction of cycloidal pendulum. (See Fig. 36.)

142. Laws of Pendulum.—

I. *Time is independent of amplitude* (subject to limitations stated above).

II. *Time varies as the square root of length.*

III. *Time varies inversely as the square root of g.*

IV. *Time is independent of mass or material of pendulum.*

143. Physical Pendulum.—The time of vibration is varied by changes in the distribution of the mass of the pendulum.

144. Center of Oscillation.—This is that point of a physical pendulum which vibrates in the same time as if it were free from all connection with the remaining particles. Its position can be determined mathematically for homogeneous bodies of regular form. In a prismatic rod suspended at one end, it is at a distance of two-thirds the length of the rod from the point of suspension. The *length* of a physical pendulum is the length of the equivalent simple pendulum, and is equal to the distance from the axis of suspension to the center of oscillation.

It can be shown that for any physical pendulum this is equal to the moment of inertia of the system relatively to the axis of suspension, divided by the product of the mass of the system into the distance from the axis of suspension to the center of gravity. (See § 145.)

The center of oscillation can also be shown to be the *center of percussion*, which is the point at which a body suspended from an axis may be struck a blow in its plane of rotation without producing any pressure upon the axis.

The center of oscillation and axis of suspension are mutually convertible; that is, the time of oscillation is the same from whichever of these points the pendulum is suspended. (See § 146, p. 65, for demonstration.)

These properties of the physical pendulum were discovered by Huygens and announced in 1673.

145. Length of Equivalent Simple Pendulum.—The length of the simple pendulum whose time of vibration is the same as that of a particular physical pendulum may be found as follows:—

Call M the mass of the given physical pendulum, and I_p its moment of inertia relative to the point of suspension P (Fig. 37). Let G be its center of gravity and O its center of oscillation when suspended from P. The distance PO is the required length, l. With the equivalent simple pendulum the mass M will be concentrated at O. Let $PG = d$.

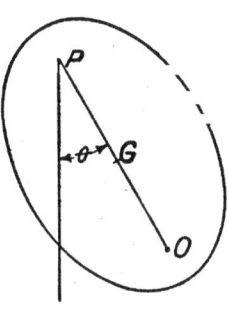

FIG. 37.

It is necessary and sufficient for equality in the periods of the physical and equivalent simple pendulum that for every value of the angle of deflection θ the angular acceleration α due to the torque produced by the weight of the pendulum shall be the same for both.

In general (§ 89, p. 38) $Tt = I\omega$ whence $Tdt = Id\omega$, and $T = I\dfrac{d\omega}{dt} = I\alpha.$

For the physical pendulum the moment of the weight, acting through G at the instant when the deflection is θ, is $T_p = Mgd \sin \theta = I_p \alpha_p$, and for the simple pendulum, $T_s = Mgl \sin \theta = I_s \alpha_s = Ml^2 \alpha_s$,

whence $\alpha_p = \dfrac{Mgd \sin \theta}{I_p}$, $\alpha_s = \dfrac{Mgl \sin \theta}{Ml^2}$. For equality of period of vibration $\alpha_p = \alpha_s$, whence $l = \dfrac{I_p}{Md}.$

The following examples will illustrate the application of the preceding principles:—

1. Position of Center of Oscillation of Prismatic Rod of length L, suspended from one end.

$$I_g = \frac{1}{12}ML^2. \quad I_p = I_g + M\,(\tfrac{1}{2}L)^2 = \tfrac{1}{3} ML^2; \quad l = \frac{I_p}{Md} = \frac{\tfrac{1}{3}ML^2}{M\tfrac{1}{2}L} = \tfrac{2}{3}\,L.$$

2. Position of Center of Oscillation of Sphere suspended by very fine wire of length h.

$$I_g = \tfrac{2}{5}Mr^2. \quad I_p = I_g + M(h + r)^2; \quad l = \frac{I_p}{M(h + r)} = h + r + \tfrac{2}{5}\frac{r^2}{h + r}.$$

146. Convertibility of Point of Suspension and Center of Oscillation. — That the time of vibration is the same from whichever point, P or O (Fig. 38) the vibrating mass is suspended follows from the fact that the length of the equivalent simple pendulum is the same in either case. Denote by k_g, k_p, k_o, the radii of gyration of the mass relatively to G, P and O respectively. The length of the equivalent simple pendulum when P is the point of suspension is

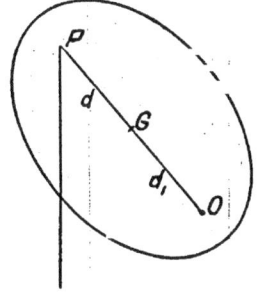

FIG. 38.

$$l_p = \frac{I_p}{Md} = \frac{Mk_p^2}{Md} = \frac{Mk_g^2 + Md^2}{Md} = \frac{k_g^2}{d} + d \ \ (1).$$

When O is the point of suspension

$$l_o = \frac{I_o}{Md_1} = \frac{Mk_o^2}{Md_1} = \frac{Mk_g^2 + Md_1^2}{Md_1} = \frac{k_g^2}{d_1} + d_1 \ (2).$$

$\dfrac{d\,d_{1}}{d_{1}} + d_{1} = d + d_{1} = l_{p}$; that is, whether the vibrating mass is suspended from P or from O the length of the equivalent simple pendulum is the same, and hence the time of vibration is the same.

147. Metronome Pendulum.—A pendulum in which a consider-able portion of the mass is situated above the axis of suspension is called a *metronome pendulum*. By vary-ing the position of a movable mass M' (Fig. 39), the time of oscillation and length of the equivalent simple pendulum can be varied within very wide limits, without increasing the dimensions of the apparatus. The metronome of Mælzel (1816) commonly used for giving time in music, is constructed on this principle, and is obviously far more convenient than the ordinary "bullet" pendulum previously employed.

Fig. 39.

It is evident that with this form of pendulum the center of oscillation lies outside of and below the mass of the pendulum it-self. Its position can be calculated by means of the formula in § 145, p. 65.

148. Determination of Length of Pendulum beating Seconds. —I. Borda's Method. (1790.)

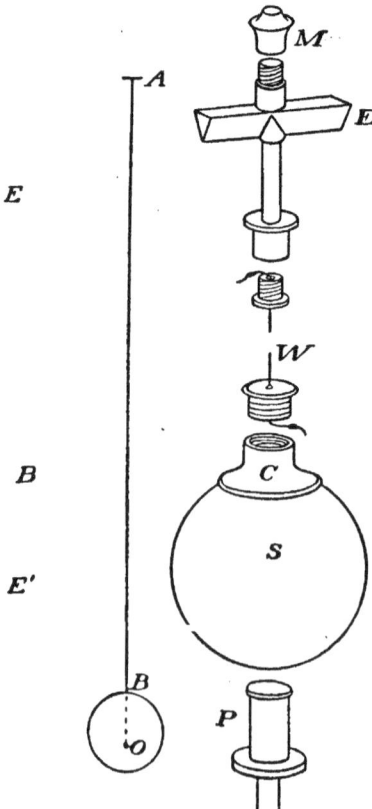

Invariable Pendulum.—Ball of platinum B (Fig. 41) is sus-pended by wire so light that it is without sensible influence on time of vibration. Length of wire (h) and radius of ball (r) are measured.

Distance of center of oscilla-tion O below center of sphere is given by analysis. It is $x = \dfrac{2}{5}\dfrac{r^{2}}{h+r}$; the length of the equiva-lent simple pendulum is therefore

$$l = h + r + \frac{2}{5}\frac{r^{2}}{h+r}.$$

The time t is determined by comparison with a clock. Then, calling l the distance, AO, l_{s} the length of the seconds pendulum,

$$t : 1 :: \sqrt{l} : \sqrt{l_{s}}.$$

The time t is determined by the method of coincidences.

Fig. 42 shows certain details of the apparatus.

positions of the weights A, B, (Fig. 40) are varied until the pendulum vibrates in the same time whether hung upon E or E'. Hence, when E is the axis of suspension, E' is the corresponding center of oscillation. The length of the equivalent simple pendulum is, therefore, the distance between E and E'. The time of vibration is measured by a clock, and the length of the seconds pendulum determined as in I.

With a reversible pendulum, if the time of vibration is approximately but not exactly, the same when suspended from either knife-edge, which is in reality the usual case in practice, it can be shown that the length of the equivalent simple pendulum is

$$l = \frac{(d_1 + d_2)\,(d_1 - d_2)}{d_1 t_1^2 - d_2 t_2^2},$$

when t_1, t_2 are the times of vibration, and d_1, d_2 the distances of the center of gravity from the two knife-edges.

Bessel (1826) used a modification of Borda's method. In this the length of the wire suspension is varied from d to d', and the corresponding times of vibration t, t' are determined.

Then $\quad l - l' = (d - d')\left(1 - \frac{2}{5}\frac{r^2}{dd'}\right)$, and $\sqrt{l} : \sqrt{l'} :: t : t'$.

Hence we can determine l and l'. d, d' = distances from axis of suspension to center of sphere. The practical advantage is that $d - d'$ is the principal term, a quantity more easily measured than either d or d'. (See Fig. 43, in which $AA' = d - d'$.)

149. Corrections.—Corrections are applied to observed results in pendulum measurements for:—

1. Amplitude of arc of vibration.
   Semi-arc used by Kater (seconds pendulum) = $1°15'$; by Mendenhall (half-seconds pendulum) = $30'$. For seconds pendulum, amplitude $48'$, correction is 1.05 seconds per day.

FIG. 43.

2. Temperatures of pendulum and of measuring rods. Lengths must be reduced to standard temperature.
   Use of "dummy" pendulum for former. "Pendulum coefficient" determined experimentally by swinging at known temperatures.

3. Reduction to vacuo: corrections for buoyancy of air, air-drag, viscosity.

Bessel devised a reversible pendulum, symmetrical in shape about its middle point and with the knife-edges equidistant therefrom. One end of the pendulum was made heavier than the other, and the knife-edges were placed so as to be approximately at corresponding centers of oscillation and suspension. With such a form the air-correction disappears in the calculation. Fig. 44, p. 69, shows Repsold's Bessel's pendulum, Fig. 45 shows Peirce's (U. S. C. S.), which is hollow, but loaded at one end.

In many cases it is preferable to swing the pendulum in an air-tight case under diminished pressure; e. g., 60 mm. at o° C, as nearly as may be, and to reduce to this exactly, the pressure correction being determined experimentally for the individual pendulum.

4. Clock correction.

5. Flexibility of support of pendulum.

6. Reduction to sea level; as g diminishes with altitude.

There are also certain other minor corrections which are necessary to ensure the greatest accuracy.

150. Results of Measurement.—The following are measured values of the length of the pendulum beating seconds and of g at certain places, according to data given in the Reports of the Coast and Geodetic Survey of the United States.

Station	Latitude	Elevation m.	Length cm.	g cm.
Washington, U. S. C. & G. S.	38°53′13″	14	99.3047	980.098
Boston, State House	42°21′33″	22	99.3335	980.382
Paris, Observatory	48°50′14″	74	99.3863	980.903
Greenwich, Observatory	51°28′40″	48	99.4134	981.171
Berlin, Observatory	52°30′16″	35	99.4230	981.265

151. Uses of Pendulum.—I. DETERMINATION OF VALUE OF g. (Huygens, Paris, 1673.)

$$t = \pi \sqrt{\frac{l}{g}} \quad \therefore \quad g = \frac{\pi^2 l}{t^2}.$$

a. Determine l_s at different stations. (1) By direct measurement of l_s (Huygens). (2) By comparison of period of invariable pendulum swung at different places; $g \propto \frac{1}{t^2}$. (Bouguer, Andes, 1737.)

Bouguer used a sphere suspended by a thread of constant length. Kater and Sabine (1820-25) employed an invariable seconds pendulum consisting of a bar terminated by a disc-shaped bob, the Indian Trigonometrical Survey (1865-75), a Repsold's Bessel inversion pendulum (Fig. 44). Von Sterneck (1880), and Mendenhall, U. S. C. & G. S. (1890) used half-seconds pendulums. Fig. 46 shows that of Mendenhall. A quarter-seconds pendulum has also been used.

b. Study of diminution of g with elevation.

(1) Determination of reduction factor to sea-level. (Bouguer, Andes, 1737.).

(2) Determination of mean density of earth ($\varDelta$). (Carlini, Alps, 1821).

Conversely, determination of mean density of mountain from $\varDelta$. (Mendenhall, Fujiyama, 1880.)

Mendenhall found density of Fuji = 2.02. Putnam (1894) found density of Pike's Peak to be δ = 2.57 ; or $\varDelta$ = 5.63, assuming δ = 2.62.

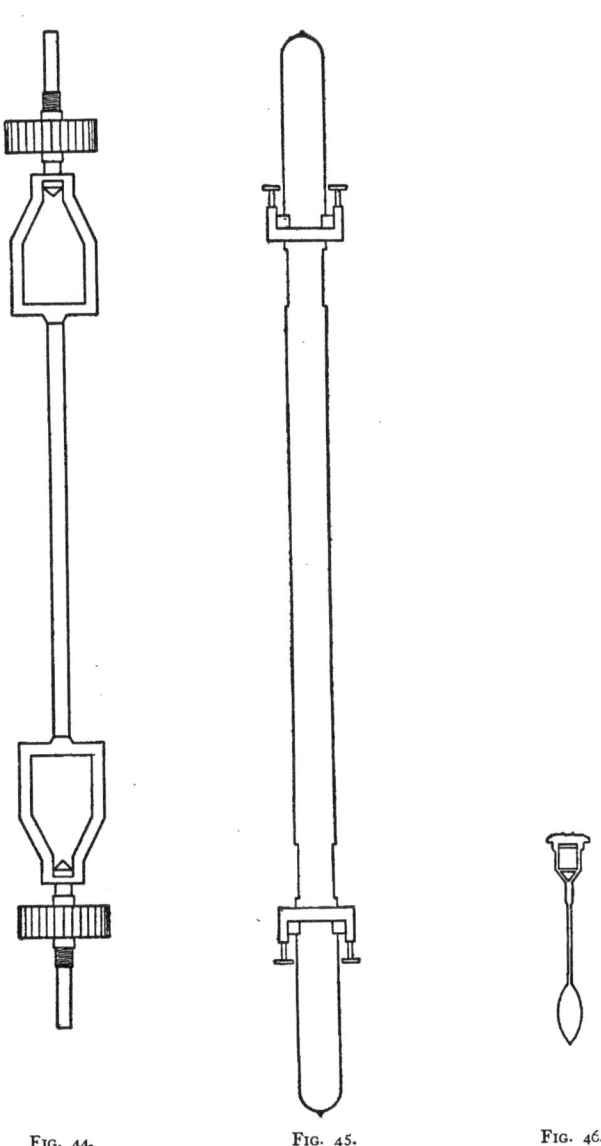

FIG. 44.   FIG. 45.   FIG. 46.

c. Study of change of *g* below surface of earth. (Airy, 1826, 1854.) Frequency of seconds pendulum at depth of 1250 feet in Harton Coal Pit increased $2\frac{1}{4}$ oscillations in 24 hours, denoting increase of *g* of 1 part in 19,200.

From this it follows that *g* at surface is greater than if there were no shell of matter above base of pit by 1 part in 14,000; from which a value of $\varDelta$ can be determined.

Von Sterneck (1883) from observations down to a depth of 1,000 meters has determined a formula representing the variation of *g* with depth.

The pendulum methods of determining $\varDelta$ are not comparable in accuracy with the methods of Cavendish and Poynting. The same is true of the determination of $\varDelta$ from observations on the deflection of the plumb line or spirit level by a mountain,—a method employed by Bouguer on Chimborazo, in 1740, and Maskelyne on Schehallien, 1774.

d. Study of local variations in *g.* Value is less than normal in mountainous regions and greater on islands and borders of continents. Observations made among the Himalayas by Indian Survey up to an elevation of 15,408 feet (Basevi). In Europe by Defforges, in United States by Defforges and Putnam.

OBSERVATIONS OF PUTNAM.

	Elevation m.	Observed value of *g* reduced to sea level cm.	Computed value of *g* cm.
Washington	14	980.101	980.087
Denver	1638	979.941	980.156
Pike's Peak	4293	979.844	980.083
Yellowstone Canyon	2386	980.369	980.605

e. *g* is independent of mass and material. (Newton, 1687; Bessel, 1832). Mass and material of bob of pendulum varied; time of vibration remains unchanged.

II. DETERMINATION OF FIGURE OF EARTH.—(Richer, 1672; Newton, 1687; Clairaut, 1743.)

From the law of variation of *g* with latitude it is possible to determine the ellipticity or oblateness of the earth. Ellipticity $\varepsilon = \dfrac{a - b}{a}$.

Clairaut showed (1743) that for a spheroid of equilibrium of small oblateness composed of concentric strata each of the same density throughout, which, presumably, is approximately true of the earth as a whole, the ellipticity is $\varepsilon = \frac{5}{2}\gamma - \zeta$ where γ is the ratio at the equator of the centrifugal force to gravity and ζ the total fractional increase of gravity from equator to pole. At a much later date (1849) Stokes showed that for such a spheroid no particular law of density need be assumed.

Clairaut also showed that the value of g in any latitude λ is given by the formula $g_\lambda = g_0 (1 + \zeta \sin^2 \lambda)$. It is easily seen that in the formula

$$\zeta = \frac{g_\lambda - g_0}{g_0 \sin^2 \lambda} = \frac{g_{90} - g_0}{g_0}.$$

It follows from these facts that the value of the polar flattening can be determined from a comparison of pendulum measurements in different latitudes.

Approximately calling $\gamma = \frac{1}{288.4}$ and $\zeta = \frac{1}{188.6}$ (its value as found experimentally by the pendulum), $\varepsilon = \frac{1}{297.1}$. The calculations leading to exact results are of course very complex. The most probable value of ε as found by the pendulum method is $\frac{1}{298.3}$ (Helmert, 1901, 1907). The geodetic method gives $\frac{1}{293.5}$ (Clarke, 1878), but this method is considered to be less trustworthy than the pendulum method owing to the comparatively small portion of the earth's surface which has been accurately triangulated. A very recent determination (1906) by the U. S. C. & G. S., using the geodetic method and based on measurements in this country gives $\varepsilon = \frac{1}{297.8}$. From a consideration of results of all the various methods which have been employed the value $\frac{1}{298.3}$ has recently been reached by Helmert (1907) as the most probable one, agreeing with the results from pendulum measurements alone.

The difference between the greatest and the least value of g is about $\frac{1}{189}$ of the minimum value, as stated above.

The actual solid dealt with in all these calculations is the *geoid, i. e.,* the spheroid whose surface would everywhere coincide with the ocean level.

FORMULÆ.—The normal value of g for any latitude λ, at sea-level, may be computed from the following formula given by Putnam (1897):—

$$g_\lambda = 978.066 (1 + 0.005243 \sin^2 \lambda).$$

The following particular values are thus obtained:—

Place	Latitude	Value of g cm.
	$0°$	978.066
	$45°$	980.630
	$90°$	983.194
Washington, U. S. C. & G. S.	$38°53'13''$	980.087
Boston, State House	$42°21'53''$	980.394
Paris, Observatory	$48°50'14''$	980.972
Greenwich, Observatory	$51°28'40''$	981.205
Berlin, Observatory	$52°30'16''$	981.294

Values of g determined by actual measurement at the five particular stations given above are stated in § 150, p. 68.

These results should be reduced to sea-level for purposes of exact comparison with the values calculated from the general formula.

A later and perhaps somewhat more precise formula than the above is given by Helmert (1901); viz.,

$$g_\lambda = 978.046 \, (1 + 0.005302 \sin^2\lambda - 0.000007 \sin^2 2\lambda).$$

The difference, however, in the results given by the two formulæ is very slight.

For an elevation h we may use the formula $g_h = g \, (1 - \dfrac{2h}{R})$, or such other as may seem preferable in any particular case.

A formula which has been widely used for the reduction to sea-level is that of Bouguer, in which for the term of correction $\dfrac{2h}{R}$ is substituted $\dfrac{2h}{R} \, (1 - \dfrac{3\delta}{4\varDelta})$. $\varDelta$ is the mean density of the earth, and δ the density of the adjacent matter lying above sea-level. Strictly considered, Bouguer's correction applies to an extended plateau. The added term is introduced in order to take account of the attraction of the mass of elevated land on which the station is situated.

VALUE OF g IN DIFFERENT LATITUDES.

(FROM EVERETT'S "C. G. S. SYSTEM.")

	Lat.	g. (measured) cm.	Elevation m.
Spitzbergen	79°50′	983.08	6
St. Petersburg	59°56′	981.90	8
Berlin	52°31′	981.27	35
London	51°31′	981.18	28
Dunkirk	51°2′	981.14	0
Paris	48°50′	980.92	70
Padua	45°24′	980.67	31
Fiume	45°19′	980.63	65
Bordeaux	44°50′	980.54	17
Boston	42°22′	980.38	22
Washington	38°53′	980.10	10
Cape Town	33°56′S	979.64	10
Port Jackson	33°52′S	979.67	
Calcutta	22°33′	978.78	6
Bombay	18°54′	978.61	11
Madras	13°4′	978.20	8
Sierra Leone	8°29′	978.18	58
Para	1°27′S	978.03	12

OCEANIC ISLANDS.

Ascension.	7°56'S	978.29	5
St. Helena	15°56'S	978.65	9
Mauritius	20°10'S	978.87	

The values of g for London, Boston and Washington have been added to Everett's table from other sources.

The following values of g were obtained by Putnam in connection with the M. I. T. party on the Peary Expedition to Greenland in 1896. A Mendenhall half-seconds pendulum was used.

	Latitude	g cm.	Elevation m.
Washington, U. S. C. & G. S.	38°53'13"	980.098	14
Sydney, C. B.	46°08'32"	980.720	11
Ashe Inlet, Hudson Strait	60°32'48"	982.105	15
Umanak, Greenland	70°40'29"	982.590	10
Niantilik, Cumberland Sound	64°53'30"	980.273	7

III. REGULATION OF CLOCKS.

The application of the pendulum to the regulation of clocks is generally accredited to Huygens (1656), although he appears to have been, in fact, anticipated by Galileo. Rival claims are advanced for Hooke, Harris and others.

A circular pendulum with a spring suspension is universally used for this purpose.

Escapement: Crown-wheel (De Vick, 1360, adopted for pendulum by Huygens, 1656); Anchor (Hooke, 1656); Dead Beat (Graham, 1715); Gravity (Mudge, 1760); Free escapement, impulse communicated through suspension springs (Riefler, 1889).

Compensation pendulums: Graham, mercurial (1721); Harrison, gridiron, brass and iron rods (1726); Reid, modification using zinc and steel rods (1812); Riefler, rod of *invar* (Guillaume's nickel-steel alloy) with short compensation tube of steel and brass (1898).

A standard clock should be kept in a constant-temperature room.

Compensation may be secured for barometric changes by the action of a magnet governed by a barometer upon a second magnet carried by the pendulum, as in the standard Greenwich Observatory clock. Or, better, the clock may be kept in an air-tight case under reduced pressure (*e. g.*, 675 *mm.*), and so protected from atmospheric variations. This was done as early as 1867 in the Berlin Observatory, and is now common. Such a clock is automatically wound by electricity.

A standard astronomical clock by Riefler at the U. S. Naval Observatory, Washington, placed in an air-tight case has run with a mean daily variation of only 0.015 seconds for a period of 3½ months. This clock possessed a nickel-steel pendulum rod and was kept in a constant temperature room.

IV. Former Standard of British Weights and Measures.

As the result of measurements in 1817 and subsequent years Kater had determined the length of the seconds pendulum at London reduced to vacuum and sea-level to be 39.1393 in. The British Weights and Measures Act of 1824 constituted as the legal standard of length for Great Britain the brass yard made by Bird in 1760, a line standard, in terms of which Kater's pendulum had been measured. It furthermore provided that in case of loss or destruction of the standard yard it should be reproduced by constructing a new standard bearing the same proportion to the length of the seconds pendulum as that borne to it by the original standard. In 1834 the Imperial standards of weights and measures, including the yard of 1760, were destroyed in consequence of the burning of the Parliament Houses. Meanwhile various sources of error had been discovered in Kater's pendulum measurements, and it became evident that a much closer approximation to the lost standard yard could be made from a comparison of the best existing copies than by employing the method of restoration set forth in the Act of 1824. A new line-standard bar of bronze (Baily's metal) was therefore constructed by the method of copying. This was constituted the legal Imperial standard yard by the Weights and Measures Act of 1855.

Under the Act of 1824 the pound (Troy of 5760 grains) was defined by reference to the weight of a cubic inch of water at 62° F. as determined by Shuckburgh.

The Imperial pound of 1855, an avoirdupois pound of 7000 grains is represented by a cylindrical mass of platinum, and is in no way referred to any other quantity.

The length of the seconds pendulum, therefore, formed the legal basis of the British system of weights and measures only from 1824 to 1855.

V. Physical Demonstration of the Rotation of the Earth.
(Foucault, 1851.)

If a pendulum could be suspended at the north pole directly in the line of the earth's axis, and set into vibration in any chosen plane, a line marking the horizontal trace of the plane of vibration in its original position would rotate with the earth in a left-handed direction at the rate of 15° per hour. Because of the permanence of the plane of vibration of the pendulum this would remain unchanged in position. Hence referred to the surface of the rotating earth the plane of vibration would appear to rotate right-handedly at the same rate.

In a lower latitude, as *e. g.*, at *B* (Fig. 47), the pendulum in the actual experiment is set into vibration in the meridian *BN*.   Somewhat later *B* will have moved to *B′*.   But the horizontal trace of the vertical plane in which the pendulum continues to move remains parallel to its original direction, and is *B′E*, which is no longer north and south in direction, but inclined to the meridian by an angle *NB′E* dependent on the latitude.   The hourly apparent rotation of the plane of vibration for latitude λ is $\theta = 15° \sin \lambda$.

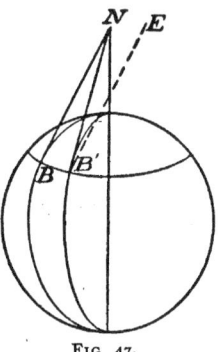

FIG. 47.

For $\lambda = 45°$, $\theta = 10° 36′$.

At the equator $\theta = 0$.

At the poles $\theta = 15°$.

A gyroscope suspended so as to move freely about a vertical axis may be used for the same purpose.   The wheel with its axis horizontal is caused to rotate with great rapidity.   If the axis is placed initially in the meridian, the revolution of the earth causes an apparent displacement of the plane of rotation of the wheel which is of the same character as that of the plane of vibration of the pendulum previously described.

In Foucault's original experiment, performed in the Pantheon, Paris, the pendulum was 220 ft. in length.   The motion was shown by causing a pointed spindle projecting below the cannon ball which formed the bob to swing through an arc of moist sand.

Permanent records were obtained at this Institute in 1876 by the use of a smoked glass plate on which the trace of the plane of vibration was registered by means of a style.

CPSIA information can be obtained
at www.ICGtesting.com
Printed in the USA
BVHW04*1024130818
524341BV00010B/178/P